FAILURE
IS NOT AN
option

About Lorraine Spurge

As the Brooklyn-born single mother of two small daughters, Lorraine Spurge began her career as a secretary and broke through the glass ceiling to reach the highest ranks of the investment banking world. There she raised more money for American businesses than any other woman on Wall Street—more than $200 billion in growth-producing capital, for companies such as MCI Communications, Time-Warner, Turner Broadcasting, Revlon, Hasbro, Mattel, Chrysler, and United Airlines, to name a few.

In 1989, Lorraine took that same dynamic spirit and created her own business, Spurge Ink! a business communications company. Her motto is simple: The More You Know, the More You're Worth.

Because Lorraine was trained on-the-job, she is capable of making complicated business issues sound simple. In her appearances as CBS 2 News' money expert, she "projects an image over the air that is warm and relaxed. She takes a subject that could be cold and off-putting and makes it entertaining and personal," says Tony Henkins, floor director for CBS 2 News. On her East Coast radio talk show, "Working with Lorraine"—which reaches 3 million listeners—Lorraine and her guests frankly answer callers' questions on all the issues facing working women today.

In 1997 alone, Lorraine appeared on more than 100 local TV and radio stations across the country, including national appearances on MSNBC, CNBC, CNN's "Business Unusual" and "It's Only Money" and Fox News' "O'Reilly Report." "One of Lorraine's best features is that she can provide smart, understandable, up-to-the-minute commentary on any breaking business news event, from an industry-wide layoff to a stock market crisis. "We'd love to have her back on a regular basis," said Don Johnson of KTVI Fox 2 in St. Louis, Missouri.

As an author, Lorraine is the editor of the award-winning *Knowledge Exchange Business Encyclopedia*. She has also written *Money Clips*, an accessible handbook of 365 ideas on how to manage your money.

As a public speaker, Lorraine has appeared in more than 25 U.S. cities in 1997 alone, including appearances at The Houston Forum, Town Hall Los Angeles, the Women's Global Business Alliance and the California Governor's Conference on Women. Lorraine serves on the Boards of George Washington University and CaP CURE, the Association for the Cure of Cancer of the Prostate.

FAILURE IS NOT AN option

How MCI Invented Competition in Telecommunications

Lorraine Spurge

Making Business Accessible

ENCINO, CALIFORNIA

Spurge Ink!
16350 Ventura Boulevard, Suite 362
Encino, California 91436

Copyright © 1998 Spurge Ink!
All rights reserved. No part of this publication may be reproduced, stored in a retrieval system, or transmitted, in any form or by any means, electronic, mechanical, photocopying, recording, or otherwise, without the prior written permission of the publisher. Printed in the United States of America. Published simultaneously in Canada.
Library of Congress Cataloging-in-Publication Data
Spurge, Lorraine, 1951–
    Failure is not an option: how MCI invented competition in telecommunications / Lorraine Spurge.
      p. cm.
    Includes index.
    ISBN 1-888232-08-0 (hardcover)
    ISBN 1-888232-41-2 (softcover)
    1. MCI (Firm)—History. 2. Telephone—United States—Long distance—History. 3. Telephone companies—United States—History. 4. Telecommunication—United States—History. 5. Competition—United States—History. I. Title.
HE8846.M375S68 1997
384.6'573—dc21   97-21903CIP

1 2 3 4 5 6 7 8 9-VA-99 98
First printing February 1998
ISBN 1-888232-08-0
ISBN 1-888232-41-2

Text Illustrations
page viii NASA; page ix MCI; page 2 PhotoDisc, Brooks Kraft, and MCI; page 3 PhotoDisc; page 6 Brooks Kraft; pages 12–14 MCI; page 16 MCI (upper and middle), PhotoDisc (lower); page 17 MCI (upper), PhotoDisc (middle), MCI (lower); pages 18–29 MCI; page 30 PhotoDisc; pages 31–32 MCI; page 33 PhotoDisc; pages 34–39 MCI; page 40 MCI (upper and middle), PhotoDisc (lower); pages 42–60 MCI; page 67 PhotoDisc; pages 70–75 MCI; page 76 PhotoDisc (upper), MCI (lower); page 78 Los Angeles Times; page 79 Beverly Rezneck; page 81 MCI; page 82 Xerox; page 83 PhotoDisc; page 84 MCI; page 90 Knowledge Exchange; page 97 PhotoDisc; pages 98–100 MCI; page 104 Lorraine Spurge; page 106 MCI; pages 112–113 MCI; pages 114–115 New York Times; pages 122–131 MCI; page 133 Financial World; page 134 MCI; page 137 PhotoDisc; pages 140–143 MCI; page 146 MCI (upper and middle), PhotoDisc (lower); pages 147, 158, 161 MCI; page 162 Los Angeles Times; pages 166–190 MCI; page 191 AP/Wide World Photo; pages 192–246 MCI; page 270 MCI: page 300 PhotoDisc; pages 301–304, 309 Knowledge Exchange.

Telecommunications Timeline
All pictures here are from the Library of Congress except for those under the following entries: Late 1870s National Park Service; 1878, 1913 PhotoDisc; 1921, 1968, early 1990s MCI; 1996 AP/Wide World Photo.

MCI Timeline
All pictures here are from MCI except for those under the following entries: Aug. 1969, May 1971 Library of Congress; Apr. 1977 Art Parts; Dec. 1978, Apr. 1983 Dec. 1991, Mar. 1992, Oct. 1994, May 1995, Jan. 1996 PhotoDisc; Dec. 1981 Xerox; Jan. 1984 New York Times; May 1984 Spots on the Spot; Jan. 1996 Microsoft; Feb. 1996 AP/Wide World Photo.

Spurge Ink! books are available at special discounts for bulk purchases by corporations, institutions, and other organizations. For more information, please contact Spurge Ink!, at: (818) 705-3740 (voice) or info@spurgeink.com (e-mail).

Dedication

*To my children, Nicole & Renee,
who have made my life worthwhile...*

Credits

Editorial

Managing Editor
Paul Murphy

Consulting Editors
J. Fred Weston, Ph.D.
Victor Tabbush, Ph.D.
Lawrance Gitman, Ph.D.

Research Director
Pamela Nelson, CFA

Rights and Permissions
Amy McCubbin

Production Manager
David Bolhuis

Research Associates
Scott Humphrey
Roseanne Landay
Greg Suess
Benjamin Tappan

Copy Editors
Paul Murphy
Janis Hunt Johnson

Proofreading
Ask Janis Editorial
& Rewrite Services

Indexer
Barbara Chamberlain

Design

Creative Director
Debra Valencia

Cover Design
Kelly Tamaki

Text Design
Frank Loose Design

Graphic Production Support
Mark Heliger
Connie Lane
Lisa Lassek
Tanya Maiboroda
Sean P. Riley
Will Soper

Acknowledgements

To the people at MCI—without whose spirit, ingenuity, and hard work this story would not be possible.

Special thanks to Bert Roberts, Jerry Taylor, Tim Price, Doug Maine, Angela Dunlap, Beverly Popek, John Zimmerman, Andrew Gruen, Stacey Leff, Pamela Montrose, Larry Harris, William Conway, Wayne English, and Philip Cantelon, without whose help I could not have completed this book.

In memory of William McGowan, with my sincere appreciation to his family—Sue Gin McGowan and Father Joe.

Contents

Prologue viii

Introduction
    The MCI-WorldCom Merger and Beyond xii

Part 1

The History of MCI and Its People 1

  1 The Molding of a Visionary 1927–1963 2

  2 MCI: The Early Years 1963–1973 16

  3 Turning to the Courts 1973–1980 40

  4 The Growth Years 1980–1984 76

  5 Expansion and New Competition 1984–1990 112

  6 The Dawn of a New Age 1990–1992 146

  7 New Directions 1992–1996 166

  8 The Future Is Calling: 1997 and Beyond 190

    Afterword 210

Part 2
A Visual History of MCI and
the Telecommunications Industry 215

Part 3
Living Lessons 247

    9 Management Strategies 248

    10 Marketing and Advertising Strategies 272

    11 Financial Strategies 302

Index 341
Glossary 351

Prologue

A crisis threatens to strand three astronauts in space. They face almost certain death, knowing they could be forever entombed in a space capsule circling the earth. The NASA project manager in Houston calls an emergency meeting with his team to brainstorm how to bring the astronauts down. He appraises the situation and says simply: "Failure is not an option."

The scene is from *Apollo 13,* a motion picture about the real-life space program. I saw the movie just about the time I began writing this book. And as I sat in the darkened theater, I couldn't help but think how those words also rang true for a business executive named Bill McGowan. In the same period as those early space flights, McGowan set his sights on another kind of challenge, but one that also involved bringing something down to earth. He wanted to beat American Telephone & Telegraph.

To understand that task, one would need to know what the telecommunications industry was like back then. American Telephone & Telegraph (AT&T) was a behemoth sustained by Federal Communications Commission (FCC) protection. It controlled 90 percent of the nation's local calls and nearly 100 percent of long-distance telephone service. The struggle was as mythic as the battle of David and Goliath—except that MCI didn't even have a slingshot.

But Bill McGowan, who grew up in a hard-scrabble Pennsylvania coal town, was not the type to fear going into battle without a weapon. After attending college on the G.I. Bill, he won a scholarship to Harvard Business School. In 1968, McGowan joined forces with entrepreneur Jack Goeken, whose fledgling company, Microwave Communications, Inc., was planning to construct microwave towers to carry radio transmissions along Route 66 from Chicago to St. Louis.

Goeken's idea was to sell communications services to truckers who needed to stay in touch with their dispatchers. McGowan, however, came on board with grander ambitions. He wanted to get into nationwide telephone service, despite—or maybe because of—the fact that the world's largest and most powerful company owned the whole business.

McGowan did what he set out to do, and the company that came to be known as MCI Communications Corporation is a thriving concern with annual revenues approaching $20 billion in 1997. The MCI story is not simply one of McGowan's personal success, however. It is also the story of a crack management and marketing team that transformed the telecommunications industry. MCI ushered in long-distance competition, triggered the 1984 breakup of Ma Bell, and paved the way for the current deregulation of the entire telecommunications field.

Bill McGowan the man and MCI the corporation, which McGowan headed until his death in 1992, embody the same creativity, ingenuity, and pure, can-do spirit that is the lifeblood of American history. Today, McGowan's successors at MCI—Bert Roberts, who at the time this book was written was the company's chairman, and Jerry Taylor, its chief executive officer—sustain that competitive spirit.

Prologue (Continued)

With the Telecommunications Act of 1996 bringing about the second breakup of a once monopolistic industry, the future is uncertain. Unresolved regulatory issues and increased competition make this time both critical and risky for MCI. But Roberts and Taylor are again positioning the company to lead the way in telecommunications diversification. In fact, MCI, under Roberts's and Taylor's leadership, announced in November 1997 that it planned to merge with WorldCom—the fourth largest long-distance company in the U.S. The move, which stunned the telecommunications industry, was hailed as a major coup that would create the first comprehensive communications company, with a substantial presence in long-distance, local, and Internet markets.

Despite the magnitude of MCI's success, the challenges it encountered along the way were much the same as those that face every start-up company. Some of MCI's setbacks would have overwhelmed even the most steadfast executives, but McGowan and company were tenacious. They did not dwell on their reversals. They learned from them and moved forward.

I first met Bill McGowan in 1983. I knew from the beginning that he was an insightful and visionary leader. But what truly struck me was his vibrancy and wit. He had a sense of fun and a lightness to his manner even when discussing the weightiest of issues.

It was a weighty issue, indeed, that prompted our first meeting. McGowan came to Michael Milken at Drexel Burnham Lambert for help in structuring a $1 billion public offering for MCI. As Milken's protégé and senior vice president of Drexel's New Issues Group for the High Yield and Convertible Bond Department, I had a hand in that colossal transaction—the largest nonutility public offering in history at the time.

Working with Milken at Drexel, I observed close-up some of the nation's biggest companies and their highest-ranking executives. Few, if any, of them intrigued or impressed me more than Bill McGowan or, for that matter, Bert Roberts and Jerry Taylor. Working on this project has given me the opportunity to get to know Bert and Jerry, and I believe that these two men are carrying McGowan's campaign to even greater conquests. For them and the rest of the MCI team, failure is never an option.

This book celebrates their tough-minded determination, optimism, and vitality, tracing the history of McGowan's MCI from its modest beginnings, and immodest ambitions, to today's corporate giant, where despite its size the entrepreneurial spark still glows.

Beyond a simple retelling of the MCI history, I have sought to take a detailed look at the imaginative marketing, creative financing, incisive corporate leadership, and employee management that supplied the muscle for MCI's battle—and that provide us with living lessons today. To delineate these lessons without impeding the flow of a truly fascinating story, I have chosen to break the book into two parts. The first part recounts the tale and introduces the characters who made it all possible. The second part takes a closer look at the lessons embedded in the struggles and successes of MCI. By calling out and analyzing in greater detail some of the specific management, marketing, and financing events that were turning points in the story, I aim to give readers a better understanding of just what drives MCI—and, more importantly, of how you can apply those defining qualities to your own business.

What I learned from MCI continues to enrich my own business experience; I promise it will do the same for yours.

Lorraine Spurge
CEO & President
Spurge Ink!

Introduction

Many analysts and reporters have mourned MCI's new alliance with WorldCom as the end of the entrepreneurial spirit that defined the company for so long. After all, if the merger is approved, the new MCI-WorldCom will be in many ways what AT&T was to the U.S.—a telecommunications giant stretching across all aspects of the phone industry. This view could not be further from the truth, however. The new alliance is absolutely consistent with MCI's original goal, which from the very beginning was to compete across all markets. As MCI CEO Jerry Taylor pointed out: "McGowan would have loved this."

A Company for the 21st Century

One of the most exciting aspects of the MCI-WorldCom alliance will be its status as a truly comprehensive communications company. Since the signing of the Telecommunications Act of 1996, a number of companies have offered rhetoric about diversifying into new markets: Long-distance powerhouses have sought to find ways into the local markets and vice versa. But the MCI-WorldCom combination will be the first to actually deliver "one-stop shopping" to its customers.

That basket of offerings will include three key components: long-distance, local, and Internet services. In long distance, MCI-WorldCom will cover roughly 25 percent of the U.S. market, closing in on AT&T with a 54 percent share (down from 62 percent in 1992). In the local markets, WorldCom brings major connections in 100 cities to the table, for one of the most substantial local-phone presences in the U.S. And in the Internet world,

MCI-WorldCom will be the strongest player in the industry: WorldCom owns the number-one Internet service provider (UUNET), and that combined with MCI's "backbone" network will make for an Internet powerhouse.

Most importantly, the new combined company will have the firepower they need to realize their separate ambitions. As a $20 billion company, MCI brings serious financial strength to the new alliance. In one gigantic move, WorldCom will leap from being an $8 billion to a $30 billion concern, with all of the financial strength to push faster growth.

MCI will also bring business experience to WorldCom—competitive lessons learned through years of hard knocks and experience. One of its most important contributions, for example, will be its unique approach to customer service. As always, MCI will be looking for ways to increase value to the customers. Take for example, MCI's call-in center, which has always recognized that customer want more than a nice voice on the phone. Customers want fast and efficient solutions to their problems. The "screen-pops" used by MCI's service representatives are an ideal example. When a customer calls the center, the representative captures the phone number and a screen pops up with all relevant information about the caller. In this way, MCI saves the customer time, gains more accurate information, and is able to solve the problem faster. These kinds of technological innovations are extremely valuable in building loyalty with a customer base and will allow the new company to grow efficiently during this time of rapid consolidation.

A Union of Upstarts

But perhaps even more important than the combination of business units will be the entrepreneurial punch of the combined companies. Prior to this announcement, MCI's most probable partner was British Telecommunications plc (BT), which offered to buy MCI for a much lower asking price beginning in November 1996. While that deal also offered important benefits in terms of global synergies, there were serious doubts about whether such two different company cultures could be successfully combined. Despite its growth beyond the U.K. monopoly it once was, British Telecom is generally perceived as being a much more traditional, cautious company than MCI.

Not so for WorldCom. In fact, WorldCom's path to the number-four position in the long-distance market rivals MCI's story in terms of sheer grit and determination. Bernard Ebbers, WorldCom's CEO, is an entrepreneur straight from the mold of MCI leaders like Bill McGowan and Bert Roberts. The press loves to play up his good-old-boy image, which is accurate (he prefers jeans to business suits and Willy Nelson to classical music), but which coexists with a shrewd, visionary approach to the telecommunications industry. Just as McGowan walked into the industry with a fresh new take on the monopolistic model, Ebbers has led WorldCom on its own independent path.

Ironically, some industry experts have even speculated that the two companies may be too much alike. Can two strong-willed personalities like Bert Roberts and Bernard Ebbers work together? And relatedly, can two such independent corporate cultures be combined without creating too much friction? These are valid concerns, but if any company is capable of harnessing explosive energy, it is surely MCI. In fact, it is exactly that kind of friction, headaches and all, which has fueled its creative success in the past. The goal is not to create a homogenous company, which in many ways implies blandness. MCI-WorldCom's diversity will be built upon rather than avoided. This strategy is nothing new to MCI, which has always embraced diversity, and believes it to be a key to its success. For example, company policy is to hire at least 50 percent from the outside, in order to encourage "new blood." People are also encouraged to moved around the company, resulting in a constantly shifting mix, and preventing people from creating completely uniform environments. In the new alliance, employees from WorldCom and MCI will be mingled with exactly this same goal of diversity.

Global Expansion from Jackson, Mississippi

Although the WorldCom alliance represents a less globe-spanning partnership than the British Telecom deal, the new company has serious worldwide ambitions. WorldCom is building local fiber optic systems in major cities around the world, including Frankfurt, London, and Paris. These high-speed lines will be used to provide voice and data services for corporate customers. WorldCom also owns one of the biggest Internet networks in Europe, a business it acquired when it bought UUNET. These worldwide efforts, combined with MCI's substantial market share in international traffic, will establish the new company as a truly global presence.

The international market offers astonishing potential. In the U.S., the telecommunications industry is viewed as a mature one, making it easy to forget how young it is from a global perspective. Even in developed European countries, the growth potential is substantial—the average person in the U.K. spends eleven minutes a day on the phone versus the average in the U.S. of twenty-two minutes. (The average in Germany is eight.) And outside of these developed countries, the potential is spectacular. China, in particular, represents an enormous source of growth, both because of its less developed state and its massive size—about four times that of the U.S. Although the Chinese telecommunications industry is growing phenomenally, installing the equivalent of one regional Bell operating company (RBOC) each year, it still cannot keep up with exploding demand. On a global basis, the telecommunications industry is still in its infant stages. Indeed, more than half of the people in the world have never even made a phone call.

Still, growth in the telecommunications industry is about much more than simple global expansion. It is about entirely new technologies and directions in communications, even here in the "mature" U.S. industry. People and businesses are fundamentally changing the way they communicate. An excellent example is presented by the change in employee communications, which is a critical part of running a successful business. Many companies now use the Internet, or corporate intranets, to update employees on company news and to exchange e-mail, and, in so doing, have created cohesive company communities. The on-line nature of this new form of communication is much more constant, for instance, than a company newsletter or management visit. And since employees are more likely to send an e-mail to management than make a phone call, intranets represent a valuable, revolutionary kind of communication between the two.

The key to this new era of global growth has been a radical shift in the attitude toward competition around the world. This shift was demonstrated perfectly in the recent World Trade Organization (WTO) agreement in which sixty-nine countries subscribed to a new worldwide measure of openness in telecommunications. Because of this agreement, MCI estimates that 95 percent of all countries outside the U.S. will be deregulated by the year 2000, versus only 17 percent today. Industry participants believe the WTO agreement to be a truly defining event, demonstrating the new emphasis on competition as a means to building a healthy telecommunications industry and infrastructure. This issue has become even more critical as many countries have watched their economies become increasingly information-based.

These changes imply huge growth for the global industry, just as they did in the U.S. Since the AT&T divestiture, the U.S. telecommunications industry has grown from about $46 billion in sales to $80 billion today. This growth is particularly surprising given that the price of a residential long-distance call has essentially been cut in half since MCI entered the industry. In fact, one of the most remarkable aspects of the U.S. transition has been the way it has benefited so many participants—customers, certainly, but also both local and long-distance suppliers. In the classic model of competitive benefit, prices dropped but have been more than offset by rising volume, leaving both consumers and producers better off. MCI foresees similar results at the worldwide level—projecting that the $700 billion global telecommunications industry will have expanded to a trillion dollars by the turn of the century.

Investor Implications

The investment community has also been intrigued by the MCI-WorldCom alliance and its sector-spanning implications. Since MCI-WorldCom may very well be the first truly comprehensive communications company since the AT&T breakup, how should it be viewed? As a telephone company, or a technology company, or both? The answer may very well be that it will require an entirely new investor category—one that reflects its true diversity—in order to accurately understand its business model and profitability. This recognition will bring new issues of valuation. Would an investor, for example, use the low price-to-earnings ratio related to the telephone industry, the higher valuation ratio of the computer industry, or a combination of both? The new alliance, and hopefully other new comprehensive communications companies that will arise in the near future, will require new measurements and attitudes from the investment community.

The new companies, because of their scope, can also offer new opportunities to intestors. True diversification will be an obvious benefit. And growth-starved U.S. telecommunications investors will find new potential in this new diverse version of the industry.

Staying Focused on the U.S.

Despite its global ambitions, the new company will focus on the U.S., which still represents nearly 40 percent of the world market; this relieves many analysts. The U.S. technological and competitive lead in the industry makes it the central player in the global telecommunications community. This dominance has been reinforced by the growth of Silicon Valley, whose concentration of software and Internet engines makes it the backbone of the Internet. In fact, the Internet has been rightly credited as one of the most important factors in reviving California's economy. Governor Pete Wilson was recently quoted as saying that the information superhighway was more important to the state than the I-5 (the largest freeway).

The geography of the United States is also still a relevant factor, as it has better access to the Eastern and Western hemispheres. Despite new technology, industry observers point out that it's still tough to lay a cable from Germany to China without crossing the U.S. Furthermore, access to capital also remains a substantial U.S. advantage. Despite progress in this area in many other countries, availability of venture capital in the U.S. is the best in the world. Certainly, MCI would not be the success story it is today had it not been for its ability to raise funds during its cash-starved growth years. Just as important, but more subjective, is the entrenched cultural attitude in many countries struggling toward free markets. Many European countries, for example, are resistant to the idea of truly competitive companies in critical industries, believing rather that jobs will be lost or that their services will become compromised. Despite the successful example of the U.S. model, these are powerful preconceptions to overcome.

Many countries also have strong nationalistic resentment of "outsider" participation in their economies, despite potential benefits to the consumer. In Mexico, for example, Telmex portrays MCI as the greedy outsider in its advertising. The television caricature is that of the greedy and vulgar American intent on robbing Mexico of its cultural identity and transporting its wealth to the U.S. Just as it was in the U.S., free market competition will always be met with attitudes that are resistant to change.

The Future of Telecommunications

The most important outcome of the MCI-WorldCom alliance will be new benefits to the consumer. Consolidation of services will be the most obvious and valuable result. Services and equipment like cellular phones, computers, and telephones will be combined, and people will be amazed that they were ever separate in the first place. In fact, as industry people like to point out, these services are essentially the same thing today—it's just that the telephone is a computer made to look like a telephone. Soon these distinctions will be unnecessary; e-mail will be available through the telephone, and voice mail through the computer.

These changes to come seem less dramatic when we consider those that have already passed. It seems almost absurd, for example, that customers once had to buy phones from the telephone company. ("Like being forced to buy toasters from the electric company," Jerry Taylor pointed out.) Across the world, customers will see new standards of availability and service, and at lower prices.

MCI Eager for Change

MCI is enthusiastic about this new environment of global change, recognizing that its ability to capitalize on shifting opportunities has always been one of its strengths. Despite being one of the "established" players in the industry now, MCI—in the new alliance—will maintain this attitude of opportunism. Indeed, given its tumultuous and successful experience in the U.S., MCI couldn't have picked a better playing field.

MCI's Story

Part 1

The History of MCI and Its People

The story has all the elements of a Hollywood movie and the drama that every little-guy saga invariably holds. But as any A-list producer will tell you, it is not a good story unless there is good character development. Similarly, companies cannot become great unless great people are behind them. MCI became extraordinarily successful because extraordinary people shaped it.

Part one of this book tells the story with an emphasis on the people and their actions: the decisions McGowan and his partners, Bert Roberts and Jerry Taylor, made that both enhanced MCI's growth and inspired its workers, and the opportunities they took advantage of without losing sight of long-term corporate goals. At the bleakest moments during MCI's early years, these dynamic people sincerely believed that failure was not an option. That belief was the driving motivation in the story you are about to read.

What emerges are pictures of larger-than-life, never-say-die personalities who are among the world's great entrepreneurs. While they can be emulated, they can never be duplicated. Any story about MCI would ring hollow were it not grounded here in the lives of the people who drove its growth and who today steer its course.

Chapter 1

The Molding of a Visionary 1927–1963

> Someone once said [of McGowan] he didn't just read books, he read libraries.
> — JERRY TAYLOR

Bill McGowan: The Molding of a Visionary

The lives of Bill McGowan and MCI Communications Corporation are inseparable in the public mind, as indeed they were to the man himself. "It's not just a job," McGowan once told *Business Week*. "It's me." But although McGowan's personality was virtually subsumed by MCI as its corporate identity, his aspirations were ignited and his management style was forged long before there was an MCI.

Coal Country Origins

Like many other young men, Bill McGowan seemed destined to walk the same path in life that his father had walked before him. Born December 10, 1927, in the small eastern Pennsylvania town of Ashley, McGowan was the son of a one-time school teacher and a railroad union organizer, both first-generation Irish immigrants.

Andrew McGowan supported his wife and five children by working as an operating engineer for the Central Railroad of New Jersey, a transporter for the region's dominant industry—coal-mining. Having started with the railroad while still a child of twelve, Andrew McGowan knew no other work than hauling coal. He understood the value of education, however, and got himself through high school by taking night courses.

McGowan and his wife, Katherine, had high hopes and expectations for their children. For Bill, they envisioned a medical degree. But hope and reality are often at odds, particularly when prevailing economic conditions dictate a different course. Typically, when workers didn't show up for their shifts, the railroad used the local teenagers to take up the slack. So like other boys at that time, fourteen-year-old Bill started hanging around the railroad yards after school, looking to pick up a paying job. With

At age fourteen, Bill McGowan started working for the railroad in various capacities from clerk to dispatcher.

the labor supply tight in a country at war, a boy showing a little initiative could get lucky and be hired full-time.

To no one's surprise, it wasn't long before young Bill was officially working for the railroad. He took on various jobs from clerk to dispatcher, sometimes working the evening shift. On more than a few occasions, he got stuck with working overnight. For another employee, that might not have been a problem, but Bill was still attending high school, and classes began about the same time the night shift ended.

Although the McGowans were a dedicated union family, Bill witnessed union practices at the railroad that he found deeply disturbing. Featherbedding, the tactic of negotiating union rules that assure more men would work on a job than were actually needed, was rampant, but McGowan understood why it occurred. Disgruntled workers who were being treated unfairly wanted to get back at their employers in the only way they knew how. From his perspective as one of the workers, he firmly believed that if a company would only treat its employees well, they in turn would treat the company the same way.

By age eighteen, McGowan appeared to be well on his way to becoming a Central Railroad lifer, except that he yearned to get a college education. While still living at home and working for the railroad at night, McGowan began attending classes at the University of Scranton in northern Pennsylvania.

His older brother, Monsignor Andrew Joseph McGowan, took a different route in life, entering the seminary at age seventeen. He remembered Bill as a natural student, possessing a passion for books and a thirst for knowledge. Bill was known for ignoring everything and everybody when he had his nose in a book, which usually consumed several hours a day.

McGowan's voracious reading habits helped him earn straight A's starting in elementary school, but he wasn't exactly an ideal student. In fact, he was invited out of parochial school in seventh grade. "Bill had a minor discipline problem," the monsignor, known informally as Father Joe, said with a grin. But it was not because he was a hell raiser. The problem was that Bill's incessant reading didn't necessarily relate to what the teacher was writing on the blackboard. Even as a young boy, he didn't like being told what to do, and his somewhat cavalier attitude toward classroom work earned him more than a few detentions.

Threatened with expulsion more than once, McGowan's neck was repeatedly saved by his mother. Mrs. McGowan, knowing how the nuns adored her son Joe, a model student, would say, "Well, if you throw Bill out, I'll have to take Joe out, too." Her strategy worked—at least for a while.

Then one fateful afternoon when Bill was being kept after school, he climbed out a window, slid down a drainpipe, and knocked over one of the nuns in the process. Shortly thereafter, Bill began attending public school.

In Father Joe's opinion Bill wanted to be thrown out. The local public school offered some interesting programs that piqued Bill's interest, and getting expelled from parochial school was Bill's only way of sampling them.

An insatiable appetite for knowledge stayed with McGowan even when he was an adult with an unrelenting work schedule. Besides having the *New York Times, Washington Post, Wall Street Journal, Financial Times,* and *Investors Daily* delivered every day, McGowan hoarded newspapers and magazines at his office. "We used to kid McGowan all the time about his reading habits," recalled Jerry Taylor, current chief executive officer of MCI. Taylor, who once even spotted McGowan reading that bible of adolescence, *Teen* magazine, added, "Someone once said he didn't just read books, he read libraries."

McGowan, however, was not a classic bookworm. He was a gregarious man with an almost childlike interest in everything and everybody. "Bill loved to talk to people from all walks of life," related his widow, Sue Ling Gin. "It was not unusual for me to come home and there would be someone sitting in my living room who had been fixing the furnace." McGowan simply enjoyed seeing the world from different perspectives and "took information from a lot of sources," she said.

This trait served McGowan well as he built MCI into a competitive force in the telecommunications industry. By staying in touch with what people from all walks of life thought, wanted,

> Someone once said [of McGowan] he didn't just read books, he read libraries.
>
> – JERRY TAYLOR

and needed, McGowan was able to match MCI's products with the diverse needs of consumers.

With just one year of college under his belt, McGowan's pursuit of higher education was interrupted by Uncle Sam. He was stationed in Germany for three years as a medic in the peacetime Army—the closest he ever got to fulfilling his parents' dream that he become a doctor.

Baker Library on the campus of the Harvard Business School.

Back home after his military service, McGowan returned to the railroad, working nights as a station clerk. He also began attending nearby King's College in Wilkes-Barre, Pennsylvania, which was closer than Scranton, on the G.I. Bill, once again immersing himself in a world of information. He devoured books about such legendary businessmen as investment banker J. P. Morgan and industrialist Alfred Sloan, who transformed General Motors into one of America's best-run companies. Through this extracurricular reading, McGowan became so enchanted with business life that he applied to do graduate work at several of the country's most prestigious business schools. He was admitted to both Wharton and Harvard—the first King's College graduate to be accepted to Harvard Business School. McGowan was leaning toward the Wharton School at the University of Pennsylvania, which was close to home and would allow him to keep his railroad job until a family friend convinced him that Harvard offered more opportunity.

The only stumbling block was money. Since the G.I. Bill would pay for just part of his first year, McGowan had to come up with the rest of his expenses. Perhaps it was McGowan's Depression-era toughness, his indomitable spirit, or just plain stubbornness, but he simply refused to give in. He sold his car, drained his savings, and, undaunted by the competition, set his sights on becoming a Baker Scholar—a distinction at Harvard that awards a full scholarship to the top 5 percent of first-year students.

McGowan was in his element in Boston, Massachusetts, and immersed himself completely. He thrived on Harvard's business case-study approach, dissecting corporate strategies and tactics. These case studies made it clear to McGowan that success or failure was related to fundamentals, not esoterica. Good marketing, attractive pricing, solid financing, aggressive sales, new products and services, low costs, high-quality operations—these basics had to be in place for a company to make it in a competitive environment. Companies did not fail because their systems were shoddy; they failed because their strategies were faulty.

By the end of his first year, McGowan had fallen in love with business—and Harvard apparently had fallen in love with him. McGowan was distinguished as a Baker Scholar and awarded a full scholarship.

During the summer after his first year in Boston, McGowan tested his increasingly sophisticated analytical skills at Shell Oil Company's New York office. For McGowan, Shell exemplified the kind of plodding, hierarchical organization that snuffed out creativity and growth. Just like the railroad employees he had known back home, people were ensnarled in the company's politics and limits. Afraid of losing their jobs, they were reluctant to express new ideas, orthodox or otherwise. McGowan vowed that if he ever had the opportunity to manage people, he would fashion an environment where employees had the freedom to express their creativity.

Later, at MCI, McGowan put his beliefs into practice by avoiding such traditional corporate hallmarks as job descriptions and written procedures. He once told an MCI executive that he would fire anybody on his management staff who developed written procedures of any kind.

You Oughta Be in Pictures

As a Baker Scholar, McGowan came to the attention of many high-level executives including Malcolm Kingsberg, president of RKO movie studio in Hollywood. Other executives warned McGowan that Kingsberg was "the meanest, nastiest bastard you'll ever meet." Nevertheless, McGowan was intrigued. For a young man, Hollywood had undeniable appeal.

Kingsberg interviewed McGowan at New York's Harvard

Club, telling him about his new film company, Magna-Theatre, which produced a revolutionary film process called Todd-AO. Todd-AO was a new wide-screen process developed by Mike Todd and used to film big-production movies like *Oklahoma* and *The Sound of Music*.

Kingsberg was looking for someone who could negotiate with theater owners and persuade them to refit their theaters and projection booths to accommodate the new equipment the process required.

McGowan jumped at the offer, and he had no trouble talking theater owners into making the risky, high-cost renovations. For a while, McGowan was riding high, making more than $40,000 a year (a substantial sum for the 1950s, when the median salary was less than $5,000), getting into other aspects of Kingsberg's operations, and rubbing elbows with movie moguls such as producers Mike Todd and George Skouras. McGowan even served as assistant producer for the movie *Oklahoma*—a distinction that he was always very proud of.

McGowan learned a lot from Kingsberg, who had a disciplined organizational style, knew what he wanted to accomplish each day, and set out to do it. On balance, though, McGowan did not like the movie business. He recognized that he wasn't particularly suited to a long-term career in an atmosphere that struck him as maddeningly unpredictable. He complained to his family that no one had any concept of costs, and money was something to be spent, not managed. This attitude ran counter not only to McGowan's education, but to his very being. Moreover, McGowan had discovered something important about himself: he didn't like working for other people. Not surprisingly, he decided that three years in Hollywood was enough.

The Entrepreneurial Spirit Takes Hold

Armed with his M.B.A. and his experience working for Kingsberg, McGowan gravitated toward the consulting business. Possessing the ability to both uncover and resolve operating and financial problems, he began rescuing companies in distress.

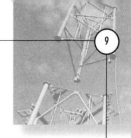

McGowan often would advise a struggling company about the kind of capital structure it should have, for example, and then help it secure the venture capital necessary to develop that structure.

His own financial arrangements usually involved obtaining a piece of the client company: the longer McGowan consulted, the larger the stake he requested. Most organizations were only too happy to oblige. "He shook up the status quo," his wife, Sue Ling Gin, remarked, herself an entrepreneur and the sole owner of the $20 million Flying Food Fare, an airline catering company. "Then the real problems would surface. He would look at those problems and take care of them."

Consulting with many different companies suited McGowan's freewheeling style and unquenchable thirst for knowledge, but he also needed an outlet for his creativity. Father Joe, who remained close to his brother throughout his life, serving as his consigliere, estimated that McGowan had a hand in creating more than a dozen companies.

Much of McGowan's success lay in his ability to quickly shift gears when necessary. His venture work with Ultrasonic Corporation of America (later known as Powertron) is a good example. The company had developed an ultrasonic beeping device that he pitched to the U.S. Navy as a shark repellent for use by pilots downed at sea. When McGowan made his presentation, complete with live sharks, to the Navy brass, however, the sharks were so fascinated by the device that they ate it! Without missing a beat, McGowan inquired whether the Navy had use for a shark aphrodisiac. Despite being one of McGowan's most exciting presentations, it was one of the few times that he didn't make the sale.

Among McGowan's earliest businesses was an investment company he started with Kingsberg's nephew and Harvard Law School graduate, Ed Cowett. COMAC (CO for Cowett, MAC for McGowan) was set up to buy into companies producing ultrasonic equipment and testing devices.

COMAC also invested in the aforementioned Powertron, which, despite the shark debacle, did have its successes, includ-

> He shook up the status quo. Then the real problems would surface. He would look at those problems and take care of them.
>
> – SUE LING GIN

ing an ultrasonic cleaner for use on small, intricate items such as jewelry. McGowan and Cowett eventually took the company public, making some money on the deal and giving McGowan experience that would prove useful more than a decade later when MCI went public.

One company on which McGowan did not make money was Geriatric Services, a nursing-home chain that he bought into in 1961 when the Kennedy administration first began considering Medicare legislation. It looked like a good opportunity at the time, but Geriatric Services finally went under because of delays in the implementation of a functioning Medicare program. This particular business may not have panned out, but McGowan's instincts were prophetic—as the country has aged, geriatrics has become a flourishing industry—and the experience only served to sharpen McGowan's understanding of regulatory matters. He recognized that unless a business could influence the regulatory process, its chances for success would depend on bureaucrats and politicians. This lesson helped prepare him for his later battle with the Federal Communications Commission (FCC) to open the door to the long-distance telephone market.

The importance of having friends in high places was also made clear to McGowan. To that end, he made a point throughout his career of cultivating certain senators and representatives, including the late Thomas P. ("Tip") O'Neill, Jr., longtime Massachusetts congressman and Speaker of the U.S. House of Representatives.

During this post-Hollywood period, McGowan made one deal after another. He took positions in both manufacturing and service companies. But it didn't take long for him to realize that tying up excessive amounts of capital in manufacturing facilities was a bad idea. Companies with relatively small amounts of fixed assets can react to marketplace changes much faster than companies with large fixed investments. MCI subsequently put McGowan's theories into practice by turning its own lack of research and development and manufacturing facilities into an advantage. Rather than depend on its own resources, MCI utilized the capacities of its numerous suppliers. The fledgling company, therefore, was able to benefit from industry innovations much more quickly and efficiently than its competitors.

Another legacy of these early years (and one that would prove to be particularly valuable in keeping MCI afloat), was McGowan's recognition of just how important management controls were to an operation. He was associated with a variety of companies, but he found that all ailing businesses had one thing in common: financial and operating procedures that were ineffective in spotting problems before they got out of hand. McGowan made sure that management controls were in place at MCI to enable it to grow without crumpling under the weight of unanticipated crises. And MCI certainly did encounter its share of challenges, but more often than not executives were able to anticipate and deal with troubles before they escalated into major crises.

One of McGowan's last ventures before joining MCI was a start-up company he called U.S. Servicator. He formed it to market a device for cars that would signal when various parts needed to be changed or fluids refilled. McGowan interested Toyota in the product and flew to Japan to meet with corporate executives there. He became disillusioned with Toyota, however, and suspected that the car manufacturer would steal his idea. He eventually sold the company.

At this point in his life, McGowan had no pressing need for a new career. He was already a millionaire at age thirty-nine. Still, he wanted to take a crack at something; he just didn't know what. Trying to sort out his life, he asked Father Joe, who had accompanied him to Japan, to extend the trip. The pair ended up traveling around the world.

When McGowan returned home, his search was at an end—although he didn't know it yet. A lawyer named John Worthington asked him if he would meet with a small group of investors searching for capital to build a series of microwave towers between Chicago and St. Louis. The group wanted to develop a private communications system for use by truckers and other commercial customers. "Why not?" McGowan thought. Enter Microwave Communications, Inc., a small, risky venture with little more than a dream and a fair amount of debt, headed by entrepreneur Jack Goeken.

Making the Connection at MCI

In the spring of 1968, McGowan, Goeken, and Goeken's partners had an exploratory meeting at a tiny Italian restaurant in Chicago. Goeken told McGowan that his company needed $35,000. McGowan wanted to know what the money was for. Goeken replied that the company owed its creditors that sum. McGowan asked, "Why only $35,000—why not get more to give MCI a little breathing room?" Goeken became defensive (the first of many times in their association), telling McGowan that MCI wanted $35,000, nothing more. McGowan knew immediately that he was dealing with technicians who, though creative, would have trouble surviving if they didn't get somebody on board with business and financial savvy.

Here was a start-up company, operating on a shoestring, that had nothing but a plan and an application pending before the FCC to act as a common carrier of communications services along a single Midwestern route. What need was there for this company or its services? The United States could boast of communications provided by the world leader, AT&T. At surface glance, MCI seemed a very long shot at best. But piquing McGowan's interest was the fact that AT&T had launched an attack on this would-be competitor—clearly the giant sensed something dangerous.

MCI Profile

John R. Worthington

John Worthington was instrumental in bringing Bill McGowan and Jack Goeken together in 1968, beginning an association with MCI that culminated in Worthington being named a senior vice president in 1979.

Although Worthington did not officially join MCI as an employee until 1971 when he became its general counsel, he had served as a member of the board of directors from its incorporation in 1968 and was made a vice president in 1975.

Worthington began his career in 1955 as an associate at the Chicago law firm now known as Jenner & Block. He became a partner in 1962, specializing in corporate financial law.

In many respects, MCI appeared to be the kind of venture that McGowan could help. Nevertheless, he hesitated. He didn't understand the industry or its potential. So in characteristic McGowan fashion he studied everything available on the subject, from newspaper and magazine clippings to books about the industry. He talked with AT&T executives, FCC regulators, and representatives of industry associations. Exhaustive preparation was typical of McGowan, but he was especially driven in the early stages of a venture to find out everything he could in the shortest time possible.

McGowan liked the idea that if he bought into MCI, he would be taking on AT&T, the world's largest and most powerful company. That kind of challenge intrigued him. He also recognized that he would have to combine his well-developed selling skills with a knowledge of the law in order to persuade the FCC regulators to allow MCI to compete against AT&T.

Bill McGowan

McGowan hadn't really worked with the government before, at least not on this scale, but his association with Ed Cowett had taught him about using the law to deal with politicians and bureaucrats. McGowan's considerable management and financial abilities would also be needed since it was apparent that none of the early MCI partners had similar experience.

Yet something still held McGowan back. When he later reflected upon his reluctance to join MCI, he concluded that he had instinctively known it would be a career-long move, and so he wanted to be sure that it was the right one.

He backed off for a short period to deliberate. Once again he and Father Joe took to the road. They went to Europe for three weeks, talking about their ambitions, their goals, and other subjects that helped clear McGowan's mind so he could come to a decision. McGowan had sought his brother's advice throughout his consulting career and re-

Jack Goeken

spected his sound judgment and uncommonly good business sense.

McGowan knew three things going in. First, whoever ran MCI would have to be able to raise a lot of money. The microwave towers, the support structure, and the people needed to put it all together would consume large quantities of cash. McGowan had been helping companies raise capital for the previous four years and had developed a large network of contacts among banks, private investors, and large institutional investors. So he felt comfortable in that role.

Second, the leaders of MCI would have to tackle the difficult regulatory issue. AT&T had built a relationship with the FCC over the years that McGowan knew would be hard to overcome. By default, the FCC had come to protect the world's largest monopoly as a means of providing the best possible telephone service for the American consumer. The few times that smaller competitors had tried to enter the market, AT&T succeeded in persuading the FCC commissioners that the start-ups didn't have the capital to last and would leave consumers holding the bag.

Knowingly or unknowingly, the FCC had been a party to squashing these fledgling challengers.

Third, and most important, MCI would have to beat the giant AT&T on its own turf—an improbable goal to say the least, given AT&T's enormous resources, which included more than $40 billion in assets. AT&T would use everything in its arsenal to try to eliminate MCI. McGowan understood that.

A crack, however, was developing in the walls of AT&T's fortress. An entrepreneur by the name of Tom Carter had invented a device he called a Carterfone that allowed two-way radios to be connected with local telephone systems. Truckers could then make phone calls without leaving their trucks. AT&T, which prohibited the connection of private equipment into its network, fought Carter with breathtaking ferocity.

McGowan decided to take the leap. He committed himself and his money to MCI, and he was undoubtedly buoyed when, shortly thereafter, the FCC ruled that Carter could use the local Bell System even though his equipment wasn't made by AT&T's Western Electric subsidiary. The ruling stunned AT&T and was a signal to entrepreneurial companies like MCI that the crack could be split wide open. For McGowan, the adventure of a lifetime was about to begin.

Bill McGowan of MCI Corporation

Chapter 2

MCI: The Early Years 1963–1973

> My brother, who is a priest, always cringes [when I say that] I've always believed that although the meek may inherit the earth, they won't increase market share.
>
> —BILL MCGOWAN

MCI: The Early Years

Jack Goeken was never considered the guy most likely to succeed. Many who knew him in his youth even might have said that he was the least likely to find success. Born in Joliet, Illinois, his father was German and his mother Swedish. Though he was the son of a Lutheran minister, Goeken was not one for the disciplines of education. In fact, he finished dead last in his high school class, barely managing to graduate. Yet Goeken was not without his merits. He had an easy manner and a quick smile, and he also happened to be good at fixing radios, which he did until he entered the Army in 1951. He was assigned to the Signal Corps where part of his training involved microwave transmission. The subject fascinated him.

When Goeken left the Army in 1953, he returned to Joliet and started Mainline Electronics, Inc., to service and sell two-way radios for General Electric. Over time, Goeken established a relationship with a small coterie of tech heads, all of whom were GE representatives who met regularly at conferences. On October 3, 1963, they decided to take their alliance a step further. Goeken, along with Leonard Barrett, Kenneth Garthe, and the brothers Donald and Nicholas Philips, organized Microwave Communications, Inc.

Chicago

This company was actually the second start-up for the team whose first venture, Communications Consultants, Inc., had sold GE radios in the Springfield, Illinois, area. It closed in less than a year after another GE representative complained to GE that Goeken's company was encroaching on his territory.

Undeterred by his first entrepreneurial failure, Goeken had a new, better, and far more complex idea: to install a point-to-point microwave-communications system from Chicago to St. Louis along Route 66. The mobile radios then used by truckers and others had a limited range of about fifteen miles. Goeken wanted to construct microwave towers every twenty-five miles that could relay radio signals from one to another. He thought the system would attract truckers, barges navigating the nearby Illinois River Waterway, and the many small businesses that dotted Route 66.

St. Louis

Jack Goeken had a vision to construct a microwave-communications system between Chicago and St. Louis along Route 66.

Private microwave systems were not so unusual in the 1960s, but the idea of one system shared by many small commercial users was revolutionary. By signing up several customers to share a private-line service that none could afford alone, Goeken believed he could offer low prices and still make money.

Goeken also unwittingly believed that AT&T would gladly welcome his venture. He reasoned that his microwave system was complementary, not competitive, to AT&T's telephone services and that MCI might even add to AT&T's revenues since it would connect to the local Bell System. The eventual realization that AT&T only wanted to bury MCI, not befriend it, came as a slap in the face to Goeken.

Although history may record Goeken as an entrepreneur, he was first and foremost a man with a passion for technology. He was full of bright ideas and the confidence and ambition to make them come to life. Goeken has been described as having the zeal of a missionary and the tenacity of a carnival barker, ready to preach the gospel of MCI to anyone who would listen.

Unlike McGowan, whose raison d'être was to beat AT&T at its own game, Goeken simply wanted to see his dream realized. But his easy Midwestern demeanor may have masked his inner resources. Behind that agreeable smile was a stubborn resolve. This combination of determination and naiveté was probably what got MCI off the ground. Otherwise, it's hard to imagine that anyone would take aim at AT&T. Not only did AT&T own the telephone business, it employed one out of every five hundred Americans and had the largest corporate law staff of any company in the United States. (When the Justice Department sued AT&T for restraint of trade in 1981, the company assigned fifty of its 930 lawyers and spent $250 million on that one case alone.)

To capitalize their venture, the original five MCI partners put up $200 each to buy stock at $20 a share. Then they subscribed to another 250 common shares at $200 a share, which amounted to $10,000 per partner. Although this stock subscription was de-

Jack Goeken did not set out to topple the AT&T empire. He simply wanted to realize his dream of creating a microwave-based communications network.

signed to raise $50,000, at this point the partners were able to come up with only $3,000, which went to cover Goeken's expenses.

By December of 1963, however, three months after MCI was incorporated, Goeken and company had enough money to file an FCC application to construct microwave towers. It did not take long for AT&T to cry foul, complaining to the FCC that MCI was unqualified to operate a microwave system and, moreover, that the system was unnecessary.

To help MCI make its case to the FCC and steer it through the miasmic regulatory maze, Goeken hired Washington lawyer Michael Bader. Bader specialized in communications law and had a particular interest in Goeken's vision. Bader's uncle, Andrew G. Haley, for whose firm Bader worked, was a pioneer in radio broadcasting law, and Bader was looking to broaden the firm's scope. In addition, Bader had just won a battle with AT&T in Texas over the operation of an intercity television system using microwave relay. That success convinced him that the telephone giant was not—despite popular opinion—invincible.

Within its first few months of existence, MCI's corporate persona of nonconformity was already taking shape. The willingness to buck conventional wisdom is an attitude that still drives MCI today. "It is the only company where you see people in a meeting who wait to see which way the wind is blowing so that they can go in the other direction," said Bob Schmetterer, a partner at Messner Vetere Berger McNamee Schmetterer Advertising Agency, which handles MCI's advertising.

Michael Bader, MCI's lawyer, specialized in communications law.

Goeken and Bader were willing to do battle with AT&T, but the wranglings with the FCC may have been more than they bargained for. Goeken later reflected that he might never have entered the fight if he had fully understood how brutal and intense it would become.

In August 1964 the FCC requested a Certificate of Convenience and Necessity from the state of Illinois; that is, Illinois had to certify that MCI's microwave towers were vital to local commerce. Unfortunately for MCI, the state wanted to conduct hearings on the matter. Fourteen months later, it finally decided to provide the certificate. The delay, however, prompted the FCC to dismiss MCI's license application. This first of many regulatory hang-ups provided a glimpse of how state agencies could waylay operations.

It was not until February 1966 that the FCC finally scheduled hearings on MCI's application. By this time, MCI's financial situation was already dire, and Goeken knew that AT&T had the intention and the power to stall the hearings further.

Hunting for Seed Money

With no cash coming in, the partners handled day-to-day operations by scraping together money from friends, associates, and vendors as well as using personal savings and credit cards (mostly Goeken's, because he was the only active partner). In fact, money was so tight that when Goeken visited Bader in Washington, he rented a car at the airport instead of taking a cab. He knew that cabbies required payment in cash.

Goeken's unbridled enthusiasm, while charming, did not sit well with potential investors. In fact, many thought he was crazy! At a meeting with one large institutional investor, Goeken recounted that the executive with whom he met offered a straitjacket in lieu of money.

This anecdote is amusing in the retelling, but it was painful for Goeken at the time. He and his partners were forced to face the fact that they were mired in the classic entrepreneurial dilemma: how to raise additional capital without losing control of the company. By February 1968, they realized that there would be no company if they didn't find an outside investor with deep pockets—and fast. Enter Bill McGowan.

It was one of those chance occurrences that changes the course of history. MCI, a company teetering on the brink of failure, had met a man whose vocabulary didn't include the word. Neither side could have envisioned just how dramatically they would succeed, or how that quality of stubborn persistence in the face of looming disaster would define their future.

But when mutual associate John Worthington brought Goeken and McGowan together for talks, the two did not become fast friends. Goeken chafed at McGowan's brash style. McGowan, in turn, saw a dreamer with little business savvy. Yet good business ideas can create unlikely partnerships.

McGowan's vision of MCI, however, was broader and more aggressive than what Goeken or his partners had in mind. McGowan wanted to restructure MCI under the franchise model, with many regional companies that would be linked together to carry nationwide telephone traffic. He reasoned that investors liked to put their money into local operations and that local investors were likely to become future customers. He also was aware that the FCC leaned toward granting radio and television licenses to local companies.

McGowan wanted to exclude the original Chicago-to-St. Louis route from the new entity because any changes in the original application would require FCC approval, which could mean additional delays.

McGowan planned to structure the network as a group of independent franchised companies, a form of operation that was becoming increasingly popular in the 1960s. Such a framework would provide all the pieces McGowan thought necessary for the puzzle: local investors, up-front cash, quick recovery of start-up costs, and the basis for rapid growth.

As the franchiser, McGowan could handle multiple FCC filings easily, reducing the time and expense of individual applications. He would also supply the franchisees with expertise in construction, staff training, and marketing. For their part, the franchisees would have to put up an initial investment of $500,000, and then additional money to ensure the completion of the regional system.

Bill McGowan and Jack Goeken put their formidable and complementary talents together to launch Microwave Communications of America (MICOM).

This umbrella organization, which McGowan called Microwave Communications of America (MICOM) was to be owned partly by McGowan and Goeken, each having 25 percent, partly by outside investors who could purchase up to 25 percent, and partly by the company, which would retain the remaining 25 percent.

The partners were swayed by McGowan's vision. In August 1968 McGowan purchased fifty shares of MCI for $35,000. That same month he spent $32.50 for incorporation papers in the state of Delaware and started Microwave Communications of America, Inc. (McGowan's $35,000 investment would prove to be a shrewd one. In 1996, 25 percent of MCI's market value came to nearly $6 billion.)

Once the new business foundation was in place, it fell to Goeken to drum up business while McGowan concentrated on one of his strengths—finding capital. With his extensive contacts on Wall Street and his network of friends and acquaintances from Harvard, it didn't take long for him to accumulate $5 million in seed money.

A Company on the Move

McGowan chose Washington, D.C., over Chicago to set up MICOM's corporate offices, a move that represented one of the company's first big expenditures. The space itself was modest, but the location was prime—just around the block from the FCC. This strategic move enabled MICOM attorneys to simply walk down the street to the FCC offices and examine AT&T's long-distance rate filings, which revealed its pricing. MCI would ultimately beat AT&T to the punch time and again because of its ability to closely track its New York–based adversary.

Between 1968 and early 1971 an additional sixteen regional carriers to cover every major market in the United States were established and incorporated, all under the MICOM banner. Much of the information gleaned from AT&T's FCC filings helped determine the most competitive locations and marketing policies for the new company.

During this growth period, McGowan grappled with important staffing decisions. He mainly was concerned with defining the kind of person best suited to move MCI forward. He didn't have the time nor the inclination to baby-sit managers. He needed people with a sense of individual responsibility and initiative, the type who could flourish in MCI's unstructured, entrepreneurial environment. McGowan figured that by hiring people who could think for themselves, he would not only get a high-quality workforce, but a loyal workforce.

Jerry Taylor, MCI's chief executive officer, was precisely that kind of person. An avionics engineer by training and a college physics teacher at the time he was hired, he came on board in 1969 as MCI's sixth employee. McGowan knew Taylor was the kind of man who responded to challenges rather than cash. Indeed, Taylor was willing to take a $7,000 pay cut to join MCI—as did many of the executives hired in those early years. Taylor, whose first job was translating technical writing into plain English for FCC filings, was also a man who could think on his feet. So, soon after joining MCI, he became the corporate firefighter, dousing fires wherever McGowan saw flare-ups.

> McGowan paid $35,000 for 25 percent of MCI in 1968. One quarter of MCI was worth almost $6 billion in 1996.

Eventually, Jerry Taylor would be known as MCI's official marketing head and in that role would guide MCI to renown as an advertising and marketing genius among American corporations. Moreover, his managerial skills were to prove indispensable to McGowan and MCI. He was the man responsible for many of the projects and strategies that laid the groundwork for MCI's continuing growth and development.

That kind of versatility is still important to Taylor and defines MCI even today. In fact, requiring employees to be flexible has helped MCI to avoid developing a traditional corporate culture. "Automorphic hiring has never occurred at MCI," Taylor explained. "We have changed and reorganized and restructured and moved people around so much, you never have people in one place long enough that they can hire others who look and talk the same way that they do."

Taylor's views are clearly grounded in his earliest years at MCI when it became apparent to him that not only did Mc-

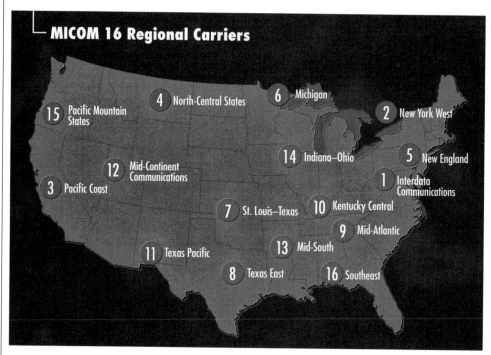

Between 1968 and 1973 MCI formed 16 regional carriers to cover every major market in the U.S., all under the MICOM banner. For maximum competitive advantage, the positions of the AT&T Bell divisions played a significant role in determining the final breakdown.

Gowan dislike corporate hierarchy, he also had no use for executive trappings like expensive luncheons and lavish quarters. Taylor remembered that in those early days his first Washington office had Formica desks and linoleum floors. When McGowan finally acquiesced to Taylor's request to get carpeting, McGowan would only approve the cheap, glued-down type.

Bill McGowan continued to disdain corporate status symbols even as the company's successes, and his own personal wealth, accumulated. Jerry Taylor also recalled a meeting that he and McGowan had with the CEO of a major corporation concerning a nationwide paging deal. "The guy was going on about his art work and his quarters at the Waldorf-Astoria Hotel. All of a sudden, Bill just got up and left," Taylor said. The CEO "continued with his stories, and it was getting a little awkward." Finally, Taylor made an excuse for McGowan's behavior and left the meeting in search of his boss. Taylor found McGowan in the office next door watching television and reading the newspaper. Knowing full well the meeting wasn't over, McGowan simply said, "Tell me when it gets interesting."

The Crucial FCC Decision

During the spring and summer of 1968, when McGowan was deciding whether or not to join MCI, the FCC was in the midst of deliberations on the company's application for a license to construct its microwave network. The application had been kicking around the commission for four and a half years at this point, and things were finally coming to a head.

Media reports speculated that the panel was split three to three and awaiting the October arrival of a newly appointed seventh member to break the tie. The swing vote would belong to the new commissioner, Democrat H. Rex Lee.

McGowan and company did not miss a beat, quickly shaping a lobbying strategy specifically aimed at winning the key vote. While Lee had served as governor of American Samoa, he had overseen the establishment of a highly regarded educational television system. Soon after Lee's appointment, the company submitted a proposal for a low-cost educational television network to the FCC, the Corporation for Public Broadcasting (CPB), and the University Communications Council. Not only did the move play to Lee's pet project, it also was designed to spotlight MCI's flexibility—and AT&T's complacency. Congress had previously requested that AT&T offer the CPB reduced rates for its educational programming. AT&T stalled on the request before reluctantly agreeing, and only after prompting from the FCC, to a limited number of reduced-rate educational hours.

Also working in MCI's favor was the changing political climate of the late 1960s. With the war in Vietnam as the catalyst, college campuses were erupting in violent protest as young Americans openly questioned their elders' beliefs about government priorities and decision making. Anything connected with the establishment provoked disdain, and few American corporations were as closely identified with the status quo as was Ma Bell. Even the new FCC regime could not ignore the shouts of "power to the people," a popular slogan that reflected the country's increasing distrust of government and regulators.

Ma Bell also had begun to show her age. AT&T's roots went back to 1877 when the first Bell company was formed. The cop-

> Requiring employees to exhibit flexibility has helped MCI to avoid developing a traditional corporate culture, which can lead to a stagnant organization where no new ideas are generated.

MCI Profile

Gerald H. Taylor

Jerry Taylor is chief executiver officer of MCI. As CEO, he oversees the global operations of one of the world's largest and fastest growing diversified communications companies. Under Taylor's direction, MCI has captured 40 percent of the growth in the U.S. long distance industry over the last five years and has become the third largest international carrier.

In 1969, Jerry Taylor, an avionics engineer in California, paid his own way to Washington, D.C., to interview with MCI. Believing he would meet with Jack Goeken, Taylor was initially miffed that McGowan showed up instead of the boss. But the twenty-six-year-old engineer was so impressed with McGowan and the company, he agreed to a starting salary of $11,000, which was a significant cut in pay.

Taylor was named to his current position in 1996 after serving as president and chief operating officer and has held a variety of senior management positions within the MCI organization, including key roles in operations, sales, and marketing.

Taylor presided over MCI's acquisition of SHL Systemhouse, a $1 billion systems integration company now known as MCI Systemhouse; he was also instrumental in engineering MCI's expansion into the $100 billion local phone service market. Other alliances with leading companies such as BT, Microsoft, Intel, and News Corp. have enabled MCI to boldly enter new markets and position itself globally while continuing to build on its core long distance strength.

Taylor has also driven MCI's expanded role as one of the world's largest Internet services providers. With alliance partner BT, MCI constructed the world's first global Internet backbone over which the two companies offered seamless, worldwide Internet and intranet services.

MCI: Failure Is Not an Option

per-based system that carried the calls had been laid down in the 1940s. Although it was once known for reliability, years of operating as a monopoly in a protected environment had finally affected AT&T's service. The Bell operating companies couldn't keep up with the demands for new services and lines, and complaints were widespread. Customers in virtually every large city across the United States—from Los Angeles to New York and from Miami to Denver—often could not get dial tones, sometimes for hours. Lines went dead for no apparent reason. Just getting a phone installed took weeks if not months. The problems in the Northeast were especially severe, prompting New York City attorneys to file a lawsuit because of "inadequate, uncertain, irresponsible" service.

Finally, after further deliberations by the reconstituted commission and nearly six years of agonized waiting by MCI, the landmark ruling was issued in August 1969. By a four-to-three vote, the FCC granted MCI permits to construct microwave towers along its Chicago-to-St. Louis route. The decision, as expected, went strictly along party lines. The FCC's four Democrats voted for MCI and its three Republicans voted against.

Reflecting the commission's ambivalence, the FCC examiner said: "Grants to MCI here might not be an unalloyed blessing. They may be an invitation to disaster; nonetheless, MCI should be given an opportunity to show that it can compete effectively."

In a subsequent ruling in 1971, the FCC also ordered that each local telephone company along the route provide MCI with

On August 14, 1969, the FCC voted 4 to 3 to award Microwave Communications, Inc. (MICOM) licenses to build a microwave communications system between Chicago and St. Louis. Celebrating this victory are, from left to right, Bill McGowan; Josiah V. Thompson II of MCI Mid-Atlantic Communications, one of the MICOM regional carriers; Alexander Buchan, a member of the MICOM Board of Directors; and Jack Goeken.

interconnection with, or access to, the local company's lines to facilitate MCI's final connection to the customer. Without that access right, MCI could not provide its service. Unlike the communications landscape of today, which is dotted with competitors in both the local and long-distance markets, state law at the time permitted only one telephone system to a city. MCI could not have built its own local system even if it had wanted to.

The victory was a heady one for MCI. McGowan would later describe the ruling ordering interconnections as both the industry's Magna Carta and its Constitution. It unbolted the door to entry in the specialized common-carrier field and spelled out the ground rules for the emerging telecommunications industry. In unbolting the door, the FCC unleashed a stampede. The commission was inundated with applications for microwave networks. When the dust finally settled, the filing frenzy had produced 1,173 applications from thirty-one different companies for 40,000 miles of microwave networks, enough to circle the globe once and still go halfway round again.

Goeken holds the "Jack the Giant Killer" trophy presented to him by Michael Bader (his Washington attorney) in celebration of the six-year battle before the FCC.

AT&T Dictates Its Terms

Not surprisingly, AT&T balked at the FCC's decision on interconnection, telling the regulatory agency that it was not technically feasible to supply all the interconnections that MCI needed. But the FCC held fast. It ordered AT&T to do whatever was necessary to give MCI access to their network.

Jack Goeken and attorney Larry Harris, who joined the company in 1972 specifically to take on the Bell System, initiated talks with AT&T about the interconnection process. Although AT&T officials were cordial enough at the first meeting, their true intentions soon became clear. Harris and Goeken traveled to AT&T's headquarters in New York City in September 1972 to begin negotiations with F. Mark Garlinghouse, AT&T's vice president for state regulatory affairs. Harris remembered that he had a sinking feeling as soon as they entered Garlinghouse's office. It was about the size of an entire floor at MCI.

"It was a nightmare," Harris recalled. Garlinghouse didn't want to negotiate. His offer, or rather ultimatum, was that MCI must not lease but buy interconnection facilities for one lump sum and then also pay a monthly or yearly maintenance charge. Goeken became agitated and started to gesticulate wildly, besieging Garlinghouse with flip charts in the mistaken belief that Garlinghouse would be more reasonable if he could only make the AT&T executive understand MCI.

Garlinghouse ended the meeting by ordering Harris and Goeken out of his office. He understood plenty—and Harris knew it. This deal would crush MCI, and as far as AT&T was concerned that was precisely the objective.

The Origins of AT&T

The name American Telephone & Telegraph was first used in 1885 to identify a New York subsidiary of the parent American Bell Telephone. The company, however, dates back to the birth of the industry in 1877. The Bell Telephone Company, which was based on Alexander Graham Bell's patents, was eventually joined with the American Speaking Telegraph Company to form American Bell in 1880.

The man most responsible for the creation of the telephone industry and its dominant force, AT&T, was Theodore Vail. It was Vail's idea to unite all the small, rival telephone systems into one network so as to make long-distance communication a reality. First as general manager of the Bell Telephone subsidiary and later as president of American Telephone & Telegraph, Vail created the regulated monopoly concept in which a company accepts government oversight in return for a dominant position. AT&T operated under that framework for more than half a century until the court-ordered divestiture in 1984.

Early operators at work in the Bell system.

> When the dust finally settled, the filing frenzy had produced 1,173 applications for 40,000 miles of micro-wave networks, enough to circle the globe once and still go halfway round again.

In an effort to straighten things out, McGowan decided to set up a private meeting with John deButts, then AT&T's chairman. McGowan recounted the epiphanic meeting to Father Joe.

First, deButts kept McGowan waiting for nearly three hours. Then McGowan was finally ushered into the office where deButts—seated in his leather swivel chair with his back to McGowan—kept him waiting for another fifteen minutes while he spoke on the phone. Finally deButts swung around to meet McGowan's gaze and said, "You know, we eat guys like you every day of the week." Astonished, McGowan replied, "You're not going to swallow this guy." The meeting was over.

From that day forward, McGowan made sure that MCI became an irritating bone in AT&T's throat that no cough or tap on the back could dislodge. McGowan later joked to his brother that he didn't get MCI off the ground, deButts did.

MCI Cultivates a High-Tech Look

With FCC approval in place, MCI decided in 1971 that it was time to acquire a little sophistication—a move that McGowan hoped would make the company appeal to potential investors and help attract new customers. Cachet came by way of CML Satellite Corporation, a joint venture formed in 1972 by MCI, Communications Satellite Corporation (COMSAT), and Lockheed Aircraft Corporation. A series of private offerings enabled MCI to invest in the $168 million satellite communications system that CML intended to build to carry messages into areas MCI's microwave system couldn't reach.

Partnering with Lockheed, a high-profile defense contractor with nearly $3 billion in revenues at the time, was an undeniable coup that gave MCI new credibility. But the venture as a whole—MCI's first strategic alliance—proved to be a miscalculation. MCI decided in 1974 to sell its interest because the business was not generating income and because it needed cash. The sale was at first blocked by the FCC, but a subsequent arrangement resulted in the sale of CML in 1975.

Admittedly, the joint venture didn't work out as the company had planned, an outcome not uncommon in either business or

Bill McGowan and MCI Board member Alexander Buchan meet with COMSAT and Lockheed officials.

private life. But what distinguishes MCI is that it wasted no time in recognizing and rectifying its misstep. One of MCI's greatest corporate strengths has always been its willingness to try new things—and to quickly abandon them if they don't succeed. The essence of the company's management style is flexibility. In fact, the joke at company headquarters is that MCI really stands for "More Changes Imminent."

Lockheed—A Name of Distinction

Renowned pilots such as Amelia Earhart and Wiley Post flew Lockheed's first airplane, the Vega, while Lockheed products like the P-38 Lightning fighter of World War II, the U-2 spy plane, and the SR-71 Blackbird are part of American military lore.

Founded in 1926, Lockheed Aircraft grew to become the nation's second-largest defense contractor before its survival was threatened in the late 1960s by a host of problems. But after government-sponsored loans saved it from bankruptcy, Lockheed recovered to develop such notable design successes as the Hubble Space Telescope and the F-117A stealth fighter.

Lockheed's 1995 merger with Martin Marietta catapulted the new Lockheed Martin into first place among U.S. defense contractors. Based in Bethesda, Maryland, the company continues to turn out such equipment as submarine-launched ballistic missiles, systems for an international space station, *Freedom*, and fuel tanks for the space shuttle.

MCI Profile

Laurence E. Harris

It took Bill McGowan two tries to get Larry Harris to join MCI. "When he first offered me the job, I thought he was insane," says Harris, who finally acquiesced in 1972 to McGowan's urgings that he direct the company's tariff and carrier relations group.

Although Harris jokes that McGowan offered him a third of what he was making, MCI's unique position in the business world thrilled him. McGowan clinched the deal by telling Harris, "You'll get to do something you'll never get to do again in your life. And you'll do something that no one else will ever do, and that's tame the Bell System."

Harris later was promoted to vice president of telephone company relations. Harris, who handled interconnection negotiations with AT&T and the FCC, was privy to all the squabbles between MCI and AT&T at the height of the confrontations.

He left MCI in 1982 to become chief of the FCC's Mass Media Bureau, and later founded and was president and chief executive officer of two paging companies, International Telecom Systems and CRICO Communications. Harris returned to MCI in 1993 as general manager of wireless communications. Now, as senior vice president of public policy for MCI Telecommunications Corporation, he is responsible for legislative and regulatory matters.

A native of upstate New York, Harris holds a bachelor's degree from Columbia University and a juris doctor degree from Georgetown University. Before joining MCI, he handled federal court negotiations for both Texas Instruments and Leasco.

"Over at AT&T, people are afraid to make mistakes," observed Jeffrey Kagan, president of Kagan Telecom Associates in Atlanta. "At MCI, people are afraid not to make mistakes."

Bob Schmetterer seconded that opinion: "I've actually heard Jerry [Taylor] say to people, 'You're not making enough mistakes.'" Schmetterer explained that too few mistakes indicates to Taylor that employees aren't "pushing far enough, fast enough—aren't being aggressive enough with their thinking and their ideas. They're being too conservative."

Taylor considers it management's job to empower and encourage employees to make decisions without fear. "I think what works best is to let people make some mistakes," he said. "Those who don't make decisions create the biggest problems for you, not the ones who make decisions and make mistakes."

Although the Lockheed alliance didn't pan out, another move made in 1971 was not a miscalculation: The corporate name was officially changed from Microwave Communications of America, Inc. to MCI Communications Corporation after it became apparent that the public was associating the word microwave with the radiation escaping from televisions and microwave ovens.

In This Deal Together—Vendors Lend a Hand

Seeking bids from suppliers to build its Chicago-to-St. Louis system, MCI found itself in a difficult situation. Raytheon Company, a Massachusetts-based firm and an early leader in the microwave equipment market, submitted a proposal for a system that, although more extensive, would cost three times the estimate of a previous bidder, the Des Moines, Iowa-based Collins Radio Company. MCI executives knew Raytheon's plan was far more serviceable, but money, as always, was an issue. Fortunately, Raytheon agreed to an irresistible financing plan: no down payment, a seven-month moratorium on payments until the Route 66 service was operational, and then stretched-out payments over a five-year period at a rate just above prime.

Raytheon took a gamble in agreeing to finance MCI's equipment. But its decision to tie its fortunes to this fast-growing newcomer clinched the deal for MCI, and ultimately gave the cash-strapped company the leg up it needed.

Important as the Raytheon deal was, MCI remained cash-poor. Although the Chicago-St. Louis microwave service was up and running by 1972, its limited commercial operations grossed only about $170,000 for the year. That meager income and the seed money McGowan had raised certainly weren't enough to keep pace with the cash outlays MCI needed for expansion. Estimates projected that the company would need $80 million between 1972 and 1977 just to construct its network.

1971 MCI logo

Even keeping up with payroll entailed some fancy footwork. McGowan would give employees their checks around 3:30 p.m. on Friday afternoon, knowing they couldn't get them into the banks until Monday, said Harris. "Then, McGowan would find money over the weekend to cover those checks."

It was MCI's vendors that again stepped in to rescue the company from desperate straits. McGowan appealed to the vendors to issue loan guarantees to a consortium of banks led by First National Bank of Chicago, with which the company was negotiating a $64 million line of credit. The loan guarantees were one part of a complicated plan conceived by MCI's new chief financial officer, Stan Scheinman, whom McGowan hired in 1971.

Under Scheinman's plan, the vendors could raise their equipment prices to MCI by up to 50 percent in return for guaranteeing the loans MCI would use to buy that equipment. The hefty price increase would compensate them for the risk they were taking in guaranteeing these loans. Almost all said yes. It was a steep price to pay for MCI, but the company wasn't in a position to negotiate a better deal. The banks were demanding the guarantees as well as insisting on a successful initial public offering of stock by MCI before they would approve the $64 million line of credit.

This financing deal did more than just provide MCI with much needed capital, however. It also solidified MCI's relationship with its vendors, which proved instrumental in their willingness to give MCI crucial support in design, engineering, and construction. "They became somewhat of a partner," related Bert Roberts, MCI's current chairman who joined the company in 1972. "And I think it gave us, as an upstart little company, an opportunity to get more response from vendors than we might have otherwise gotten."

From the vendors' perspective, backing MCI provided an added bonus: another company besides AT&T that could buy their products and services. On the other hand, there were a few vendors that wouldn't touch MCI because they were afraid AT&T would take its business elsewhere. This situation would reoccur when MCI was seeking financing. Many investment houses chose not to work with MCI for fear of alienating Ma Bell.

In hindsight, the financing plan appears brilliant, but it was also an extremely risky proposition. MCI had to have the line of credit, but the banks' approval depended on the company first drumming up some serious cash by selling stock in the public market. By involving the vendors in guaranteeing the bank credit, MCI was putting vendor faith on the line. If the initial public offering (IPO) did not succeed, not only would the credit evaporate, the vendors' confidence and business would too.

For the all-important task of taking the company public, MCI hired the Wall Street investment banking firm of Blyth & Company, Inc. (eventually absorbed into PaineWebber). Blyth scheduled an offering of three million common shares, or about 40 percent of the company, for the early summer of 1972.

The IPO

June 22, 1972, was the day of reckoning, and it was fraught with anxiety. That morning, when

M C I **Profile**

Stan Scheinman

Stan Scheinman joined MCI in 1971 as senior vice president and chief financial officer to help finance Microwave Communications of America, the corporation McGowan formed to consolidate MCI's operations.

Before joining MCI, Scheinman was vice president of administration for PepsiCo. When his division was shut down, he moved to the finance division of Revlon. Scheinman oversaw MCI's initial public offering in 1972, selling stock in the open market to help finance network construction. He also arranged for vendor financing at a crucial time in MCI's history.

His imaginative use of financing and leasing arrangements helped MCI to obtain the capital needed to stay afloat. He resigned from the company in 1975.

> **Then and Now: MCI and Netscape**
>
> A present day analogy to MCI's offering in 1972 is Netscape's IPO in 1995. Both companies had little but a new telecommunications concept to attract investors, and both companies were stunned by the investor response.
>
> | | **MCI**
 June 1972 | **NETSCAPE**
 August 1995 |
> |---|---|---|
> | **Offerings** | 3.3 million shares | 5 million shares |
> | **Proceeds** | $30.2 million | $140 million |
> | **Use of proceeds** | Financing construction of telephone microwave | Financing construction of Navigator system on the Internet |
>
> MCI's offering was expanded to satisfy investor demand, making the offering one of the largest IPOs in history. Netscape's experience was even more dramatic: the shares, which were sold at $28 in the offering, were driven as high as $78 that first day, before closing at $57—more than double the offering price.

the IPO was to become effective, Hurricane Agnes ripped through Washington, flooding roads and preventing the Securities and Exchange Commission official responsible for approving the offering from getting to his office. The 10 o'clock opening of the stock exchange came and went, but MCI stock couldn't begin trading without SEC approval. When the official finally made it in, he promptly approved the offering.

The response to the IPO probably took even the ever-bullish McGowan by surprise. The three million common shares were fully subscribed at $10 a share, and the underwriter added another 300,000 shares from its over-allotment option to cover unexpected demand for the new issue. The offering netted MCI a little more than $30 million and gave it a total market value of $120 million, satisfying the terms of the bank credit line received a week earlier. It was the second-largest IPO by a development-stage company in U.S. history. (COMSAT, with whom MCI partnered to form CML, held the IPO record at the time.)

Besides reflecting MCI's growing credibility in the eyes of the public, the offering was successful in part because of its timing. The year 1972 was a good one for going public—$3.3 billion was raised through IPOs, a 72 percent increase from the previous year. Stock values were moving up as well. On January 11, 1973, the Dow Jones reached a new high of 1,051.70.

With financing in place, MCI was finally in a position to build the network that McGowan had visualized.

Life in the Very Fast Lane: Building a Nationwide Network

The euphoria produced by the IPO was short-lived as MCI realized the task it faced: building a nationwide telephone network in less than a year. Considering that it had taken AT&T forty years to build its own system, the job seemed impossible. But then again, seemingly impossible jobs were the ones that most appealed to MCI.

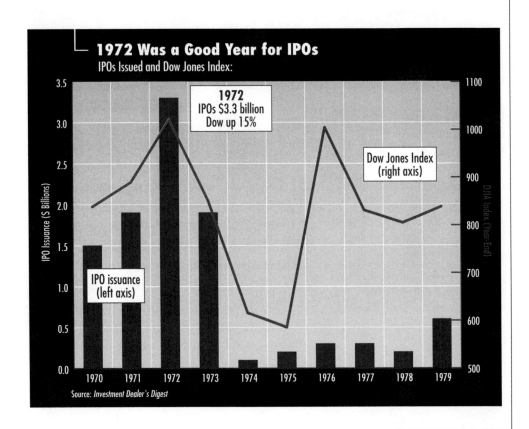

1972 Was a Good Year for IPOs
IPOs Issued and Dow Jones Index:

1972: IPOs $3.3 billion, Dow up 15%

Source: Investment Dealer's Digest

Responsibility for building MCI's new microwave network was assigned to Thomas Leming, a microwave engineer who had taught in the Navy's radar school during World War II. Fortuitously for MCI, the intercity private-line network that Leming was working on in St. Louis was abruptly canceled. When McGowan came to woo Leming, the timing was perfect.

Like McGowan, Leming had attended college under the G.I. Bill. And also like McGowan, he was smart, strong-willed, hardworking, and not afraid to express his opinion. Dubbed "the white tornado" because of his shock of white hair, Leming was the type of guy who worked overtime to stay in touch with his employees and keep abreast of their needs. He had the strength of character that inspired respect and admiration in others.

Just how much inspiration Leming could engender was tested in the fall of 1972, when he gave his construction managers an astonishing deadline: Build the routes from Chicago to New York and from Chicago to Dallas by the following summer. Leming's people thought he was crazy, but they were willing to try.

AT&T had built its copper-cable system in the 1940s using standard procedure: map out the system, do a site survey, purchase or lease the land, build an access road, run power lines, level the site, and then fine-tune everything. Leming's people, to put it mildly, moved at a much faster and less conventional pace.

Field crews charged from site to site, pouring concrete and bolting down steel towers at top speed. Details were to be worked out later as the crews drove themselves to finish site construction. Often too involved and then too exhausted to go to the nearest town and find a

MCI Profile

Thomas L. Leming

Thomas Leming was senior vice president for Continental Telephone before coming to MCI Communications Corporation in 1971 to build its national telecommunications network.

The original Chicago-to-St. Louis microwave transmission route was completed on time and on budget through a marathon effort orchestrated by Leming. By 1982, he had planned, constructed, and made fully operational a state-of-the-art international network that carried long-distance calls via satellite, microwaves, and fiber optic cables. He retired from MCI in 1985.

Leming helped develop microwave technology for Western Electric, Motorola, Lenkurt Electronics, Collins Radio, and Hughes Aircraft.

Work crew constructs a microwave tower in the race to complete the system on time.

motel, workers slept in their cars, which they euphemistically dubbed the Camaro Hotel. Prefabricated radio-equipment shelters were trucked to the sites to save even more time.

These field crews symbolized everything McGowan wanted MCI to be: quick, innovative, adaptable, hardworking, and bureaucracy- and pretense-free. Resolutely slogging their way along the two routes to Texas and New York, the crews were getting the job done. Then, in the summer of 1973, MCI's cash shortages began to hamper their progress.

The earlier $80 million cost estimate had been seriously understated. A more sophisticated analysis of construction and equipment costs put the price tag for building the network at $180 million. What's more, orders from the FCC to the contrary, AT&T continued to delay providing timely interconnection to MCI, thus keeping it from generating needed revenue even as it linked more cities to the system. By the end of 1973, revenues from limited commercial operations in the cities where MCI had been able to obtain interconnections were running at only about $100,000 a month, while company expenses were well over $2 million.

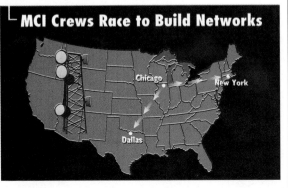

Chapter 3

Turning to the Courts 1973—1980

> Two years from now I hope to reach a level of business equal to Bell's annual bad debt provision.
>
> — BILL MCGOWAN

Turning to the Courts

When today's business pundits reflect on MCI's astonishing rise, they point to its litigious nature as a pivotal factor in its success. MCI eventually won millions of dollars in damages from AT&T that helped finance its operations, and McGowan himself often referred to MCI as a "law firm with an antenna on the top."

This comment, however, should not be misconstrued. MCI did not sue AT&T because it thought it could make money without actually earning it. MCI sued AT&T because McGowan was damned mad. It wasn't just that AT&T was trying to clobber MCI; McGowan could handle that. It was the company's methods that offended his sense of fair play. "He didn't mind if they played hard or rough. He didn't mind if they used every trick in the book (and they did) if they did it honestly," explained Larry Harris. "But what he most resented about them was that they didn't do it fairly." Harris recalled that one of AT&T's most underhanded tricks was to disconnect its circuits on MCI's eastern corridor route.

Bill McGowan before a Senate subcommittee recounting AT&T's violations of antitrust laws.

MCI had appealed to Judge Clarence Newcomer of the U.S District Court in Philadelphia concerning the interconnection issue, and in December 1973 he ordered AT&T to comply. The company did so, but soon appealed the decision. To AT&T's elation, the appeals court vacated Newcomer's order.

Before the FCC could intervene on MCI's behalf—which it soon did—AT&T thought it would teach this pushy upstart a lesson. On a Friday afternoon, just as McGowan and company were to embark on a weekend management seminar, AT&T pulled the plug on MCI's circuits between New York and Washington. Without warning, the data line at a major Washington department store went dead, and other corporate customers were similarly disconnected. It's not hard to imagine the chaos that ensued as one after another of MCI's customers lost their private-line service. And of course, MCI's apparent inability to guarantee connections did nothing to endear it to the few customers it had.

McGowan may have been scared, but in front of his troops, he did not blink. He wanted to get right into a strategy session. "I thought, what are we doing, arranging the deck chairs on the Titanic?" Harris remembered. Harris—and probably most of the other managers—was convinced that MCI was sunk.

For McGowan, this latest affront was like rubbing salt in a wound after his rancorous meeting with AT&T's chairman. McGowan became more determined than ever to fulfill the promise he had made to himself that fateful day. He would not be swallowed by AT&T.

Haggling over the interconnection issue, McGowan realized, was not getting MCI anywhere. In an interim agreement, AT&T had dropped the proviso that MCI buy its interconnection facilities, but its rates were still ridiculously high. They would soon run MCI aground if McGowan didn't do something to prevent it. He believed litigation was the answer. As usual, however, money was a problem. MCI had money either to build a network or to litigate, not both. McGowan called a meeting of his closest advisers, including Kenneth Cox, a bright regulatory lawyer and former FCC commissioner whom McGowan had recruited from that office in 1970. Knowing it was an uphill battle, Cox suggested that MCI take a backdoor approach that included sending material detailing AT&T's obstinate stance to the Antitrust Division of the Department of Justice.

MCI also began a campaign in Congress that resulted in a congressional hearing at which

MCI Profile

Kenneth A. Cox

Kenneth Cox was MCI's senior vice president in charge of regulatory affairs from his hiring in 1970 until he retired in 1987.

A former FCC commissioner who had been appointed by President John F. Kennedy, Cox's FCC post was about to expire when Michael Bader first approached him.

Although Cox felt uncomfortable moving to a company that didn't even have a product yet, Cox found MCI (at the time MICOM) interesting enough to accept an offer where he would work half-time for MCI and half-time for Bader's well-established law firm.

During his tenure, Cox participated in the regulatory battles of interconnection, equal access, and the antitrust suits against AT&T.

McGowan waxed eloquent and passionate. "AT&T, despite the wishes of the user, despite the orders of the FCC, and despite the laws of the country is attempting to maintain a monopoly in intercity communications," he said. McGowan proposed to Congress that Ma Bell be split into separate entities—one for long-distance service and others for local service.

MCI carried its cause to anyone who would listen including shareholders, customers, and friends. McGowan sent out twenty-five thousand letters beseeching people to help MCI defeat AT&T's anti-competitive campaign.

Although MCI's backdoor plan was stirring up sympathy from both the public and the government, this course of action alone could do little to ease the pressures on MCI's business. So in March 1974, MCI began its frontal attack when it filed a civil suit against AT&T in the U.S. District Court for northern Illinois, charging violations of the Sherman Antitrust Act. The Chicago law firm of Jenner & Block, where MCI's general counsel, John Worthington, was once a partner, agreed to take the case.

Searching for Revenues: Execunet Saves the Company

At about the same time, just as MCI needed more cash to expand its network, the lending banks were carefully scrutinizing every dollar the company spent. In early 1974 the country was in a recession. Credit was tight and banks were more inclined to call in loans than to hand them out to an upstart that had the temerity to sue the world's largest and most powerful corporation.

To free up whatever cash it could, MCI had already sold a subsidiary, Spectrum Analysis & Frequency Planning, Inc., which performed microwave path analysis and frequency protection, to Collins Radio for $175,000. Six months later it made plans to sell

1980s TAC dialer. Revenue received from Execunet, a shared private-line, low-cost service, sustained the company through lean times.

its interest in CML Satellite, but the deal didn't materialize right away. Finally, in July 1975 MCI sold CML to IBM for $2.5 million.

MCI was on the spot. Most of its revenue came from 4K Plus, which provided private-line voice, data, teleprinter, and facsimile service to companies in the fifteen cities then hooked up to the MCI network. Though pleased with its reception in the corporate community, 4K Plus clearly could not generate cash quickly enough to meet the spiraling demand. MCI had to develop new services that would produce the highest possible revenues at the lowest possible cost.

It fell to Bert Roberts, then vice president of operations, to find the solution. Roberts, an electrical engineer from Johns Hopkins University, had come to MCI two years earlier after a friend suggested that he meet Bill McGowan. At the time, the thirty-one-year-old Roberts had been offered a new job that included a title and a raise. But before accepting, he decided to take his friend's advice. Roberts caught up with McGowan in an airport (McGowan was on one of his frequent money hunts), and the two hit it off. Roberts accepted McGowan's offer: no title, no raise, lots of risk. He had been bitten by the MCI bug.

To keep costs to a minimum, Roberts looked for a service that would utilize MCI's existing network, mesh with the skills of its employees, require minimal working capital and new equipment, and be up and running quickly.

Roberts, with his engineer's grasp of technology, turned to a switching device recently discovered by Jack Goeken that allowed customers to route long-distance telephone calls over the least expensive available line, be it MCI or AT&T. Throughout the summer and fall of 1974, MCI tested the product, known as a WATSbox, at Rohm & Haas, Inc., a pharmaceuticals company based in Philadelphia.

The system could also break out and analyze each long-distance call by number and length and assign each call a special account billing code. An additional option, known as remote access, allowed a subscriber to call into a central number and be connected to the long-distance line without a credit card or an operator, offering significant savings.

Roberts's team took the idea a step further by developing a shared private-line service that could operate between the cities that MCI served. It was expected to appeal to small- and medium-sized businesses wanting the lowest-cost long-distance service available. It was called Executives' Network, which was soon shortened to Execunet.

Making a call on Execunet required a touch-tone telephone at a time when the vast majority of Americans had standard rotary-dial phones, but was relatively simple otherwise. Subscribers punched in an access code to hook up to the shared Execunet line, waited for a tone from the switch, punched in an authorization number, received another tone, and then punched in the area code and number.

The only hitch was that customers had to punch in up to twenty-two numbers to make a phone call. A pain in the neck? Yes. But Roberts thought the timing was right. In 1974 the country was in a deep recession and businesses were desperate to

In its early, cash-strapped stages, MCI was caught in the classic start-up dilemma—lots of money going out with little coming in. MCI was spending money rapidly to build its network, but it earned no revenues from operations until 1974.

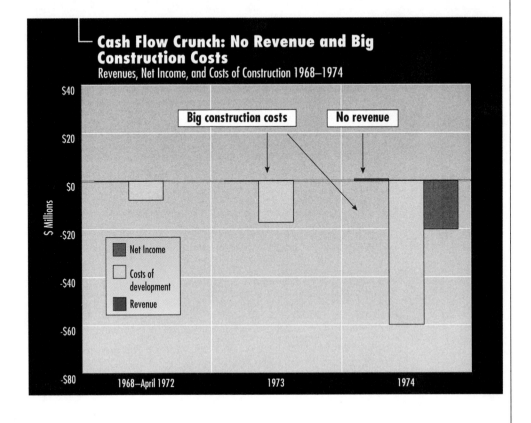

Cash Flow Crunch: No Revenue and Big Construction Costs
Revenues, Net Income, and Costs of Construction 1968–1974

M C I Profile

Bert C. Roberts, Jr.

Bert Roberts, MCI's chairman, joined the company's sales and marketing department in 1972 at the age of thirty-one. He describes his move to the company as a "right angle change." Having spent more than ten years in the computer industry, first at Westinghouse and then at Leasco (a computer time-sharing company), Roberts traded in the security of working with well-known, established companies to follow the siren song of an exciting, risky, entrepreneurial upstart.

McGowan's hiring of Roberts proved to be a watershed event for MCI, as Roberts's creativity and grasp of MCI operations have contributed mightily to the company's success. It was Bert Roberts who came up with the idea for Execunet, which rescued the company when bankruptcy threatened in the mid-1970s. He was also a key player in MCI's expansion abroad, first with the 1982 acquisition of Western Union International and later in its pivotal relationship with British Telecommunications. Not to be overlooked is Roberts's skill as a negotiator in tariff proceedings and his organizational talents during the 1985 decentralization and subsequent restructuring, both of which proved crucial at important junctures in MCI's history.

Rising steadily through the ranks and gaining knowledge of virtually every aspect of MCI operations, Roberts became president and chief operating officer in 1985 and chief executive officer in 1991. After McGowan's death in 1992, he was also named chairman.

Although Roberts had been handpicked by McGowan, some observers worried that his more sedate style might translate to a dull, too cautious MCI. Roberts quickly laid such fears to rest, however, pushing the company to widen its horizons so as to make the idea of convergence in the telecommunications industry a reality. Now widely acknowledged as a leader who is firmly focused on the future, Roberts has even advised the Clinton administration on telecommunications policies designed to encourage the development of the information superhighway.

Under Roberts's command, MCI has become the world's third largest carrier of international traffic, build the world's fastest and largest Internet network and established technology alliances with industry leaders such as Intel and Microsoft. In addition, Avantel, MCI's joint venture in Mexico, became the first company to receive a license to compete in Mexico's recently deregulated long-distance market.

Roberts's personal commitment to better education has helped forge MCI's pioneering efforts to advance the use of communications technology in the education arena and communities at large. Recent programs include: NET Vote '96, a national online effort to involve students in the 1996 presidential campaign; Net Day '96, an initiative designed to provide Internet access to thousands of California schools; and CyberEd, a campaign in support of a White House initiative to bring hardware, training, connectivity and content to more than 400 schools in 15 designated "empowerment zones" across the country.

An admitted technophile, one of the first things Roberts does every morning before breakfast is to flip open his laptop PC to check his e-mail—and he's notorious for sending messages after midnight and before dawn. Having been quoted as saying, "You can reach me anywhere in the world, any time day or night," he does, in fact, receive up to one hundred messages a day, from low-level MCI employees, fellow executives, and MCI customers.

But e-mail is more than Roberts's favorite medium. It's a metaphor for the kind of culture he continues to cultivate at MCI. He wants revved-up, type-A people who don't need face-to-face supervision or coffee-machine shop talk, people willing to work odd hours in odd places, but always accessible through e-mail—just like the chairman himself.

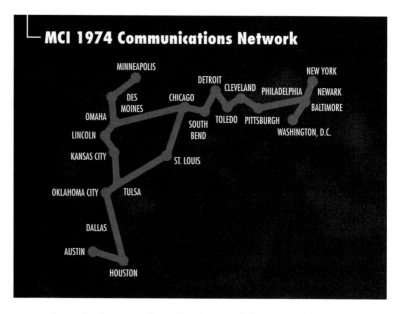

In 1974 most of MCI's revenue came from 4K Plus, which serviced companies in the fifteen cities then hooked up to MCI network. Though it was well-received, 4K Plus clearly could not generate cash quickly enough to meet growing demands.

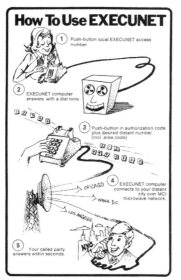

strengthen the bottom line. He doubted they would mind dialing a few extra numbers to save money. The detailed report of calls provided by the service would also be an attractive feature to companies that wanted to track their communications expenses as well as build databases on their customers.

Convinced that Execunet could produce the revenue MCI so badly needed, Roberts went to McGowan for approval in June 1974. At first, McGowan balked. He feared that revising MCI's tariff at the FCC—by law, all services MCI offered had to be covered by MCI's FCC tariff—would open a new regulatory battle with the commission and AT&T. McGowan just wanted a revenue-producing service with a quick start-up, not another wrangle that might jeopardize its antitrust case. But McGowan gave in a few months later, and filed and won approval for a new tariff for the service.

From the beginning Roberts thought that Execunet was ideal

for lawyers who required detailed information about calls for billing. Where better to test the market, he reasoned, than Washington, D.C.? Results of the pilot run were disappointing, however, and that particular market never did materialize. But Roberts refused to give up. He pushed to expand the service to Dallas.

Jerry Taylor, who by then was running MCI's operations in Texas, agreed with Roberts and thought that Execunet had tremendous potential. His gut instinct told him the service could work between Houston and Dallas where there were no federal or state regulations on intrastate long-distance telephone calls. Taylor pleaded with McGowan to let them try Execunet in Texas. "McGowan really didn't have much interest in doing it," Taylor recalls. "It was Bert [Roberts] who gave me support."

By January 1975 Taylor had assembled a hungry sales force that worked solely on commission. "I couldn't afford [salaried] salespeople, so I hired straight commission salespeople," Taylor recalled, "and these guys knew how to pound on doors and sell." They didn't need, or want, to know anything about telephones other than how to use them. In fact, most of them were copy-machine sales reps. Ripping out clumps of yellow pages, they began cold calling—with great results. Texans loved the idea of cheaper calls and began signing up by the thousands.

Execunet's resounding success pleased Taylor to no end. He loved this brash, can-do sales approach, particularly since AT&T had never telemarketed. It was yet another example of MCI shattering industry conventions to get a leg up on AT&T. As for Roberts, he couldn't help

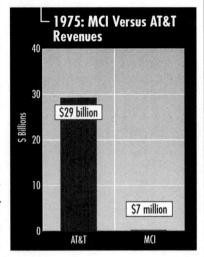

1973–1980 ☏ Turning to the Courts

but feel both elation and relief. His first big project for MCI was obviously a winner even if it had stumbled coming out of the gate.

Within two months Execunet's revenues were surpassing its expenses, and by September 1975 MCI's Dallas One, a speedier and more technically advanced switch to carry Execunet calls, began operating. The new hardware used in Dallas One enabled MCI to route traffic from Kansas City, Chicago, Washington, and New York through Dallas while also increasing the number of channels available for expansion. All indications were that Execunet would eventually be a gold mine for MCI. Moreover, MCI had established itself as a technological leader in the increasingly competitive communications industry.

Parting Ways and New Beginnings

As MCI had been tested over the years, so too had the relationship between Jack Goeken and Bill McGowan. Each new challenge highlighted their vast differences. Goeken's dream was to branch out and develop new technologies such as high-speed faxes. McGowan, on the other hand, was a builder and a fierce competitor who loved the game of business. While always intrigued by technological innovation, McGowan's primary focus was on building a company. Their different approaches to growth made a break inevitable.

The kind of rift that developed between Goeken and McGowan is not uncommon in the realm of entrepreneurism. Often, the person with the idea and the fire to launch a company is not the person best suited to lead in its further development. And it's not uncommon for the pioneer to be edged out by the builder. MCI needed both Goeken and McGowan, but by July 1974 Goeken finally decided that he had had enough. He took his 1.5 million shares—worth more than $3 million at that time—and, without ceremony, quit. (Goeken went on to start three other companies: Airfone, Inc., FTD [Florists Transworld Delivery], and InFlight Phone Corp.) The parting was not easy for Goeken because MCI had been his baby, but his role in the

MCI Profile

John D. Goeken

Goeken left the MCI story early on, but remained a major player in the telecommunications industry. After leaving MCI he went on to spearhead the development of the air-to-ground telephone industry. He began by forming Airfone, Inc. in 1976, which gave air travelers the first opportunity to make phone calls from the air. The idea proved to be a major success and he sold it to GTE in 1986. (It now operates under GTE AirFone). He stayed with the company for three years after the sale, but left to start a competing company after a dispute with management.

This step might have been tricky for anyone other than Goeken. GTE AirFone held a monopoly on the business at the time, but to Goeken—nicknamed "Jack the Giant Killer"—this state of affairs only made the challenge more interesting. He broke GTE's monopoly by convincing the FCC to allow frequencies to competitors, and even won $15 million in a lawsuit against GTE in the bargain. He started his own company, InFlight Phone Corp., in 1989. In an interview with the L.A. Times, his daughter commented, "Jack's always hated monopolies, unless they were his own."

True to form, Goeken pushed InFlight Phone toward new technological innovations, making it the first air-to-ground company to offer digital transmission. This distinction gave airplane passengers access to such new services as fax transmission and even transmission from laptop computers. But Goeken, as always, kept moving and sold the company in 1994.

Next up—the Goeken Group, which develops and markets specialized applications within the wireless communications industry. A Goeken Group personal security monitor, for example, will notify a center in case of trouble; another device will store medical records at a central database for transmission in case of an accident. The group received a boost when a new corporate investor recognized the value of his ideas and bought 5 percent of the company. In a way this last transaction brought Goeken full circle. The new investor was MCI.

company had been diminishing for some time.

Shortly after Goeken quit, MCI's lending banks put pressure on McGowan to find an experienced manager skilled at running a large corporation. That man was V. Orville Wright, who came on board in May 1975. Previously an executive at IBM, RCA, and Xerox, Wright was named president and chief operating officer of MCI by December.

The two men were a study in the attraction of opposites. Wright, the professional manager, perfectly complemented McGowan, the entrepreneur. A Midwesterner from Kansas, Wright had been a Navy officer, while McGowan, an Easterner from Pennsylvania, had been an Army enlisted man. Where Wright was an administrator, a product of corporate organization and structure, McGowan deplored established procedures, as evidenced by his well-known threat to fire anyone caught writing a company systems handbook.

If McGowan was combative and confrontational, Wright was the picture of composure and stability. Where McGowan was aggressive and impulsive, Wright was measured and conservative. McGowan took risks; Wright worked to turn them into viable strategies. Dissimilar though they were, these two men blended together to build a stronger MCI. The *Washington Post* aptly described Wright as the man who kept "MCI's feet on the ground, while Chairman William McGowan reached for the stars."

When asked to comment on his relationship with McGowan, Wright said, "Some people have described it as yin and yang.... He's the spokesman who loves podiums. And I've been the inside man who worked with employees, who did the nitty-gritty work."

Perhaps one of the keys to how well these two men seemed to mesh despite their differences can be found in Wright's words, "I've been the inside man who worked with employees." Both Orville Wright and Bill McGowan loved people. Wright, in fact, was instrumental in developing the company's first employee awards program. And not only did he stress the importance of people power at MCI, he was generally interested in his employees personally and took pains to monitor their well-being

> My brother, who is a priest, always cringes (when I say that) I've always believed that although the meek may inherit the earth, they won't increase market share.
>
> – BILL MCGOWAN

MCI  Profile

V. Orville Wright

Orville Wright joined MCI Communications Corporation as senior vice president in May 1975, and was named president and chief operating officer the following December.

He continued in that position until his retirement in 1985 when he was succeeded by Bert Roberts. Wright came out of retirement in 1987 to serve as acting CEO after McGowan had his first heart attack, then retired again in 1990 at the age of sixty-eight. He remained vice chairman of the board and a consultant to the company until 1991.

Wright, who was with MCI in the early years when things were most precarious, recalled the 1970s as a time when "you didn't know if you were going to make it. We had regulatory battles and we wondered if we could attract the big customers. Would they give part of their business to a young company without much of a network?"

Wright was satisfied to work behind the scenes handling daily operations while McGowan, always visible, was out looking for money and dealing with the courts and the FCC.

Before coming to MCI, Wright worked for IBM for twenty years, starting as a sales trainee in 1950. He was director of marketing for IBM's federal government accounts and then director of its Systems and Technology division. He moved to RCA in 1970 to become head of its Computer Systems Development division and later served as director of marketing for Amdahl Corporation and vice president of business development for Xerox Corporation.

Wright graduated from the University of Kansas before serving with the U.S. Navy for seven years.

and performance. One of Wright's former staff members recalled how much delight he took in meeting and talking with each award-winning employee.

Furthermore, Wright and McGowan were in fundamental accord about the future of MCI, and both understood the immense dedication and effort necessary to make the company successful. Thus, they sought compromise—or at least Wright did.

Wright agreed that the company would be run in an entrepreneurial fashion as long as possible with few restraining rules and regulations. McGowan agreed to more structure once it was needed. But change comes hard for some people and, as Wright discovered, MCI's entrepreneurial period was to last far longer than he expected.

MCI and the FCC Battle Over Execunet

Just as Orville Wright was getting settled in at MCI, the wave the company had been riding for a few months due to Execunet came crashing down. Responding to complaints from AT&T, the FCC in July 1975 ordered MCI to stop offering Execunet. It might as well have asked MCI to shut its doors. Unwilling to yield, MCI did what it was quickly learning to do best—litigate. It filed suit with the U.S. Court of Appeals and won a stay of the order, allowing the company to continue selling Execunet. This salvo marked, however, the start of more than three years of litigation over the service.

Execunet's monthly revenue, already about $36,000, was expected to grow dramatically. MCI estimated that Execunet could generate revenue of $40,000 per month per current WATSbox, and that a new, larger switching system would produce up to $170,000 per month. With the company envisioning the addition of more and more WATSboxes in more and more cities, MCI hunkered down to fight for what it believed was a potential cash cow. Even McGowan, who had never been completely sold on Execunet, rallied to the cause. Both he and Roberts forcefully argued before the FCC to save Execunet. Neither man was willing to quit, particularly when Execunet could supply the cash MCI desperately needed to complete its network, pay down its bank loans, and ensure the company's survival.

Meanwhile, the banks gradually had increased MCI's line of credit until it stood at $80 million. But at this figure they drew the line. Despite the rosy projections for Execunet and a June ruling by the Supreme Court that in effect ordered AT&T to provide MCI with interconnections for authorized services, the lenders were adamant that they would not provide another cent. So MCI turned to the capital markets.

Unfortunately, the negative publicity and the uncertainty provoked by the endless round of FCC rulings, lawsuits, appeals, and court-ordered stays had taken its toll on the price of MCI's common stock. From the $10-a-share IPO price in June of 1972, the stock had plummeted to $2.38 by mid-1975, and stubbornly hung around the $2 level throughout the year. Hating to enter the market with the stock so depressed but having no other viable alternatives, MCI went ahead in November with an oft-de-

The negative publicity generated by the regulatory proceedings took its toll on MCI's share price. It fell from its IPO price of $10 to $2 3/8 by the middle of 1975. Desperate for cash, MCI had no choice but to sell stock at rock-bottom prices.

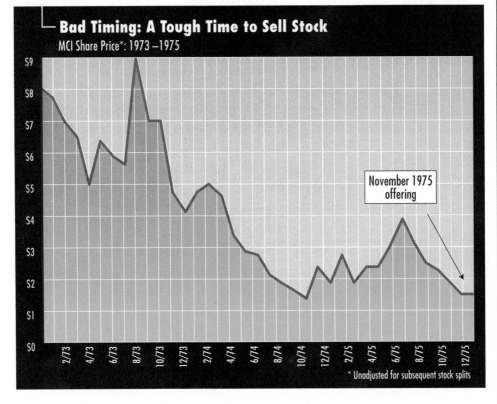

Bad Timing: A Tough Time to Sell Stock
MCI Share Price*: 1973–1975

November 1975 offering

* Unadjusted for subsequent stock splits

layed secondary offering of MCI common stock and warrants.

McGowan, like any owner, was always reluctant to sell equity no matter what the price was, but this giveaway was particularly devastating to him. The $8.5 million the offering produced was crucial to MCI's survival, however, and he knew it.

Despite MCI's compelling arguments on behalf of Execunet, the FCC, after a court-ordered review, formally rejected Execunet once again in May 1976, proclaiming it was not the private-line service MCI had been authorized to sell. Only the earlier appeals court stay allowed the service to continue pending further litigation.

When the Copying Machines Are Repossessed, You...

It appeared that McGowan's Irish luck had run out. Jerry Taylor recalled that things hit rock-bottom when, after failing to pay the lease fees, MCI's copying machines were repossessed. But MCI management held firm in its belief that the company could succeed and were determined to get through the crisis. They just had to convince employees not to panic and abandon what appeared to be a sinking ship. "We had to say things like, 'We're going to get through this crisis, but this stuff is going to happen around you. One thing we'll never do is put equipment and other deals ahead of paying your paychecks,'" Taylor recounted. And the company was true to its word.

But although MCI always held fast to its commitment to people before equipment, the bleakest days during the struggle over Execunet did force some layoffs. Technically in default and unable to pay interest on its loan, holding down expenses was the highest priority. Operations were consolidated, travel and overtime limited, and reimbursement of entertainment expenses was discontinued. Despite stringent cost-cutting measures, however, some 250 people were let go in the winter of 1975 and 1976, and fifty more were laid off in June. With an employee count of between 650 and 750 people at the time, this number amounted to more than a third of the workforce.

MCI knew Execunet was a winner if only the company could

stay in business long enough to gain court approval of the service—and if it could somehow keep its creditors at bay. MCI's banks had good reason to be nervous. Having softened from their earlier position of "not another cent," the banks had agreed to further incremental lending until MCI owed them almost $95 million by early 1976. If they declared MCI to be in default, which technically it was, most of the money loaned to the company would be lost.

McGowan, who could find humor in even the most desperate situations, joked that the banks were caught in the typical lender's dilemma. "If I owe you a hundred dollars, that's my problem," he said. "If I owe you a hundred million dollars, that's your problem."

Loan Schedule Stretch-Out Gives MCI a Breather

Apprehensive, yet unwilling to pull the plug, the banks agreed to defer MCI's interest payments for a few months more. In return for their patience, they demanded that MCI hire an experienced chief financial officer.

The company had been trying to fill that vacancy since June 1975 when Stan Scheinman resigned amid a flurry of innuendo that the banks were unhappy with his performance. Scheinman's job had been a thankless one that mainly involved informing the banks of MCI's typically bleak financial picture. He was an asset to the company, but the banks' scapegoat nonetheless.

After Scheinman's resignation, MCI's hardworking and dedicated controller, Richard Uhl, served as de facto CFO (although McGowan officially assumed the title), but the banks thought Uhl didn't have "enough gray hair." Not surprisingly, few executives with top-level operating experience were willing to take on a high-risk, low-paying position with a struggling company.

MCI approached Wayne English, an experienced finance executive with substantial background in lease financing and raising capital. English's résumé was extensive. He worked his way up the ladder at Ford Motor Company and then moved on to Pullman Company, an automotive and industrial parts manufac-

MCI Profile

Wayne English

Wayne English joined MCI Communications Corporation in February 1976 as vice president, treasurer, and chief financial officer. He retired from MCI in 1984 at the age of sixty-one, but continued as a member of the board of directors until 1992.

English was the executive who first contacted Michael Milken at Drexel Burnham Lambert in 1981 to help raise capital, which led to MCI's financial independence after almost twenty years of being capital-poor.

Prior to joining MCI, Mr. English was an executive-level financial officer for a number of large corporations. He started his career in 1946 with the public accounting firm of Lybrand, Ross Bros. & Montgomery (later Coopers & Lybrand), where he was a senior accountant.

From 1949 to 1955, Mr. English served as a financial analyst for Ford Motor Company, including two years as Financial Officer of Aeroquip Corporation, where he served for eleven years. In 1966 he joined Pullman Incorporated as vice president for finance. Two years later Mr. English was appointed senior vice president of finance for Trans World Airlines where he also served on the board of TWA's wholly owned subsidiary, Hilton International Corporation.

In 1971 Hallmark Cards, Inc. retained Mr. English as its executive vice president, responsible for the company's financial management and the administration of its real estate subsidiary, Crown Center Redevelopment Corporation.

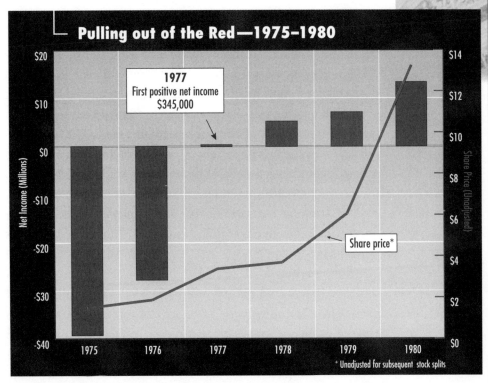

turer. He went on to serve as CFO for Trans World Airlines (TWA) and then as executive vice president responsible for finance at Hallmark Cards. Furthermore, English was tough enough to stand up to McGowan.

Before taking the job, English asked to talk with MCI's lenders. They stonily informed him that MCI owed the consortium $95 million, that MCI had never repaid any principal or interest on the loans, and that MCI was surviving only through their good graces.

Surprisingly, this frosty encounter did not discourage English. He was captivated rather by MCI and by the challenges it represented. "I suppose," English said, "I get the same kick from raising money that a salesman gets from landing a big account." The banks gave their nod of approval. Not only did English have an impressive résumé and demeanor, he had a full head of white hair.

The appointment of English as CFO, additional cost-cutting, and increased revenues from Execunet led the banks in the fall of 1976 to defer interest payments, which came to more than $8 million annually until 1977. With operations and investments eating up $60 million a year, this deferment was no small matter.

This good news was followed by something even better: MCI posted its first monthly profit in September, followed by its first profitable quarter in December, which was true cause for celebration. At the end of the fiscal year in March 1977, MCI reported its first yearly profit as well with net income of $345,000. At last, significant amounts of money were starting to flow in to the struggling company, and mostly thanks to Execunet.

The banking consortium, having witnessed some positive performance, further agreed in January 1977 to revise the terms of MCI's credit agreement, which originally called for ten equal quarterly payments of principal between 1977 and 1979, a schedule that would have severely shackled MCI's ability to expand. The new agreement extended the payment schedule to 1984 in exchange for which MCI agreed to meet certain income and cash flow targets. MCI now had the breathing room it needed to complete construction of its network, increase its operating revenues, put most of its financial house in order, search out new ways to expand services, and pay down debt. Moreover, tensions between MCI and its banks finally began to ease as the lenders started to see a possibility of getting their money back.

The Courts Back MCI

The Supreme Court of the United States

In July 1977, precisely two years after the FCC first banned Execunet, the U.S. Court of Appeals voted unanimously to allow MCI to offer the switched intercity-communications service without restrictions. Both the FCC and AT&T appealed the decision to the U.S. Supreme Court, which six months later declined to review the case. MCI had won. Execunet was now legal, and a whole new world of profit opportunities opened up for the company.

In April of the following year MCI won yet another huge victory in the appeals court. For the second time the court ordered

AT&T to provide MCI with local interconnections for its Execunet service. This time, however, AT&T chose to willingly comply, resolving instead to take a different tack in its efforts to thwart MCI. Promising to file new tariffs to cover the Execunet connections, AT&T indicated that the new rates would be "significantly higher."

Wrangling continued. As in the past, each victory seemed to lead only to new problems, further delays, and additional legal costs for MCI. First, AT&T claimed that the courts had ordered it to provide local interconnections for MCI's private-line service only, not for Execunet. Then, after yet another appeals court ruling specifically ordered AT&T to make Execunet interconnections available, AT&T agreed to provide them—at triple the current prices! MCI executives appealed to the FCC, which agreed with MCI and urged AT&T to avoid further legal squabbles and negotiate a reasonable settlement for all interconnection fees.

Obviously, AT&T's strategy hinged on attrition. AT&T intended to defeat MCI by draining away its resources, frustrating its customers, and demoralizing its people. Time and again Ma Bell struck, and time and again MCI survived the challenge. The court victories aroused real jubilation in the MCI work force. Each blow that was deflected made employees even more devoted to ensuring MCI's success.

David and Goliath Negotiate

In September 1978 David and Goliath sat down to resolve some issues. Executives from MCI and AT&T met in Washington to negotiate the interconnection fees. The negotiations revolved around AT&T's proposal to triple interconnection rates over the ensuing eighteen months—the net effect of the new ENFIA tariff it had filed the previous May. ENFIA stands for Exchange Network Facilities for Interstate Access, and was the acronym used to signify the connection between local telephone systems owned by Ma Bell and specialized long-distance carriers such as MCI. MCI opposed paying any ENFIA tariff at all, arguing that the local Bell operating companies already charged Execunet cus-

Employee Dedication

An astonishing example of the unselfish dedication MCI employees display for their company occurred during the summer of 1977 when New York City was engulfed in a blackout. As soon as he saw the lights begin to flicker around dinnertime, an MCI engineer realized the software for MCI's switch, located on the twenty-fourth floor of the J. C. Penney Building at 1633 Broadway, would crash during a blackout, despite automatic auxiliary power, if commercial power were not restored within a few hours.

The engineer actually raced to the building and climbed the twenty-four flights of stairs, only to find the office doors locked. He ran back downstairs, convinced a security guard to manually open the elevator doors, and climbed up through the elevator shaft. By prying open the elevator doors, he was able to gain entrance to the switch room where he properly shut down the switch before any permanent damage could occur.

In addition twelve employees from Cleveland, concerned about the crisis and eager to help out, had taken the first plane to New York when they heard about the blackout.

---

tomers for local telephone service, and they should not be allowed to receive double compensation for this service. Furthermore, MCI deeply resented AT&T's arbitrary filing of the ENFIA tariff without any negotiations whatsoever.

AT&T viewed the proposed new rates as simply an effort to recoup part of the cost of providing long-distance service. But to the specialized carriers, the crux of the issue was that if AT&T was making money selling interconnections at the regular business line rate, why should charges have to rise so sharply under the ENFIA new tariff? MCI considered the new tariff just another attempt to put it out of business. The action, its lawyers claimed, was "irrefutable evidence of a conspiracy between AT&T and its associated companies to restrain competition and to unlawfully create a monopoly."

At the urging of the FCC, nearly every long-distance company involved in telecommunications joined the negotiations. Representatives from GTE, International Telephone & Telegraph, Southern Pacific Communications (the parent of Sprint), and a half-dozen other companies and telephone cooperatives attended, but it was evident that AT&T and MCI were the major players.

The chief negotiators, Bert Roberts and William Stump, vice presidents of MCI and AT&T respectively, wanted a compromise, not an impasse. Roberts suggested that the specialized carriers pay local-network access charges tied to their levels of market share and expressed as a percentage of the base amount paid by AT&T's Long Lines unit. As long-distance revenues rose, so would the percentage paid to the local Bell operating companies. AT&T liked the concept, and the details were quickly hammered out.

As will be seen later, however, this resolution of the interconnection issue would open the door to a new controversy—equal access (the process that would allow customers to choose their long-distance carrier). In the years ahead, as the specialized carriers grew and their local access fees approached 100 percent of the base amount paid by Long Lines, these carriers would begin crying foul and demanding a level playing field. They would end up getting a whole new ball game.

One participant in the ENFIA negotiations had wryly called the agreement "peace in our time," but the war wasn't over, not by a long shot.

Network Links Added Nationwide

Despite the turmoil of the endless court appearances and corporate maneuvering, early in 1978 MCI began to expand its network in earnest. With approval from its lending banks, MCI had purchased a microwave system from Western Tele-Communications, Inc. (WTCI) for $6.5 million. The populous western market was an obvious target for MCI's expansion, but high construction costs kept the company from building its own systems there. Since 1975, it had leased capacity on the WTCI system, which connected Phoenix, Tucson, San Diego, and Los Angeles.

In June 1978 with revenue from Execunet continuing to increase, MCI bought the system outright. The purchase also included a second route between Denver and San Francisco, and

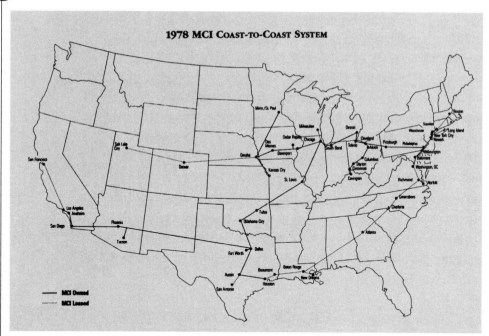

1978 MCI Coast-to-Coast System

The map above indicates areas served by MCI after its western expansion through the purchase of WTCI. The acquired WTCI system is highlighted.

gave MCI access to the West Coast. The cost of acquiring the system was significantly below the cost of building it. Most importantly, MCI was finally able to link Los Angeles and San Francisco, thus completing the West Coast network.

MCI was aggressively moving into other regions as well: between New Orleans and Norfolk, Virginia, in the Southeast; Chicago to Milwaukee in the Midwest; and upstate New York and major New England cities in the Northeast. Senior MCI executives realized that the best way to increase sales was to expand the network. The company was beginning to enjoy what McGowan called the "beneficent circle": growth producing profits that generate more growth that produces more profits.

As the number of cities in the network steadily increased, the economic impact was readily apparent. In the years between 1968 and 1974, the company's revenues from communications services, mainly 4K Plus, had totaled $728,000. In 1975 alone when Execunet came onto the scene, total revenue soared to $7 million. Even with the legal and regulatory straitjacket confining Execunet, revenue continued to surge, reaching $28 million in 1976, and $74 million by 1978.

By the end of fiscal 1979, Execunet had almost three hundred salespeople in fifty cities. The number of customers for the service had grown to fifteen thousand, and new installations were running close to one thousand per month. In 1979 MCI's profit nearly doubled to $7 million. Its success was rewarded in the stock market as its market capitalization jumped from $63 to $140 million.

The Public Is Responsive to MCI Preferred Stock Offering

With revenue and profit climbing, investor confidence in MCI was on the rebound as well, giving Wayne English a chance to launch his plan for untangling MCI's web of bank debt. English's strategy involved making use of bank credit on an interim basis to finish building the network, then tapping into the public markets for money to repay the banks. If the formula was successful, the process would be repeated.

The first step involved a public offering in December 1978 of 1.2 million shares of convertible preferred stock that paid an annual dividend of $2.64 per share—a 10.6 percent yield. Priced at $25 per share, the sale netted $28.6 million.

Convertible preferred stock offered advantages both to MCI and to investors. Although preferred is senior to common stock

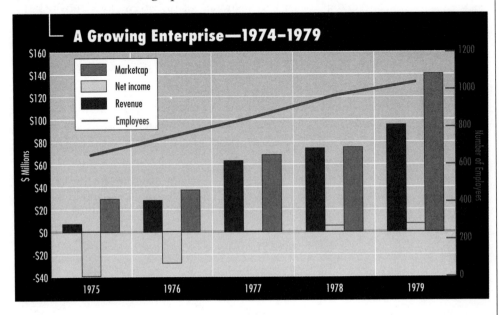

December 1978 Preferred Stock Offering: The Details

On December 12, 1978, approximately 1.2 million preferred shares were issued for $25 each, yielding $28.6 million after fees. Each preferred share was convertible after April 4, 1979, into 5.7 shares of MCI common stock at $4.38 per share, a premium of almost 17 percent over its then-current price of $3.75, which meant that investors would make money on the conversion only if MCI's stock rose 17 percent or more. (All share prices and amounts are unadjusted for subsequent stock splits.) MCI had the right to redeem the preferred shares on or after January 1, 1980, at prices declining from $27.64 in 1980 to $25 in 1986, plus accrued dividends.

> The 80 percent deduction can make a big difference to the investor. Assuming a 40 percent tax rate and a 10 percent gross return, the deduction will boost the after-tax return from 6 percent to 9.2 percent.

in the event of a liquidation and pays the buyer a fixed rate of interest like a bond, it is not as punitive as debt. So missing a dividend payment doesn't put the issuing company in default. For a company with hefty cash needs and a still-precarious financial position, this was an important consideration. Moreover, the convertible feature allowed MCI to sell stock at a higher price in the future than the price at which it was then trading, which made it potentially very lucrative for a company with high expectations.

Finally, the tax considerations involved were positive for both the issuer and the buyer. MCI was not yet earning enough money to move it into the ranks of taxpayers, so the fact that the dividends it paid out were not deductible was of no consequence. It could take advantage of the benefits offered by a convertible preferred issue without having to worry about the drawbacks. But for corporate investors, preferred stock provided a very attractive tax advantage: 80 percent of the dividends received were not subject to taxes.

MCI's share price did gain substantially, standing at $7.81 by the end of April 1979 (when investors could convert them), nearly 80 percent above the conversion price. In effect the $25 paid for each preferred share was worth $44.63 (5.7 times $7.81) at the conversion date only four months later.

As part of English's complex financing arrangement, MCI earmarked $11.5 million of the proceeds from the preferred offering to prepay bank debt and an additional $4 million was used to cancel warrants the banks held to purchase slightly over four million shares of MCI stock. (The banks had received warrants as an equity kicker in 1975 in return for their willingness to defer interest payments.) Since MCI had only about 20 million shares outstanding at the time, a 4-million-share position would have represented substantial ownership by the banks had they converted the warrants. MCI was pleased to eliminate that possibility.

Commenting about the financing, English joked that "they [the banks] were the first at the feeding trough whenever there was money to be had." But the package was extremely beneficial to MCI as well, particularly since canceling the warrants prevented future dilution of equity at low prices. The convertible preferred offering also enabled the company to reduce its debt, thereby lowering soaring interest costs. (MCI was then paying almost 4 percent above the prime rate on its bank loan. With interest rates in the late 1970s hitting new double-digit highs every month, this made for very expensive interest payments.)

There was general agreement that MCI emerged from the offering a stronger company, and investors gave MCI a vote of confidence in a most satisfactory way—they traded the stock. In 1978 MCI's stock was the ninth most actively traded security on the NASDAQ over-the-counter market, accounting for volume of about 16 million shares.

The ever-present appetite for new funds to expand the network and grow the company brought MCI back into the capital markets in September 1979. A public offering of senior cumulative convertible preferred stock raised $69.5 million, $2 million more than had been

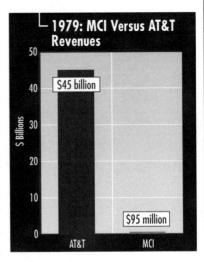

1979: MCI Versus AT&T Revenues

AT&T: $45 billion
MCI: $95 million

September 1979 Preferred Stock Offering: The Details

MCI raised $69.5 million (less underwriting fees) by selling 4.95 million shares of preferred stock at $15 a share. Each share of preferred stock paid a dividend of $1.80—for a 12 percent yield—and was convertible into 1.5 shares of common stock at $10 a share. (All share prices and amounts are unadjusted for subsequent stock splits.) Since the common stock was selling at $6.50 on the day of the offering, the conversion price represented a 54 percent premium to the market. MCI called the preferred stock for conversion in May 1981 at a price of $16.50 a share. The underwriters were Shearson Loeb Rhoades and Allen & Company.

raised in all previous public offerings combined.

English had originally thought he could sell about two million shares, but underwriters Shearson Loeb Rhoades and Allen & Company suggested doubling that number. When their analysis proved correct, MCI was able to pour half the offering's proceeds into network expansion, allocating the rest to buying back outstanding warrants, paying off part of the credit agreement with the banks, and increasing working capital.

The extra money from the increased offering began to ease some of the pressure on English. For a few months at least he didn't have to respond to one cash crisis after another in a struggle to keep the company afloat. Now he could turn his attention to long-range strategic planning and attempt to anticipate needs and arrange financing in advance.

The primary goal of MCI's financial plan was to pay down bank debt to reduce borrowing costs, which were considerable with the prime interest rate at 15 percent. In addition English wanted to firmly establish the company in the capital markets to enable MCI to finance the massive amounts of cash required to build its network.

MCI Launches Residential Service

As the company continued to utilize the capital markets, management believed the time was right to bare its teeth to AT&T. Although its first antitrust suit against the company was not yet settled, MCI filed another one in April 1979, which sought $3 billion in damages for AT&T's actions in restraint of trade from 1975 to 1978.

The cost of such a suit could have been financially prohibitive, but MCI struck a deal with its law firm to cut its fee in exchange for a percentage of the damage awards. As it had done earlier with its vendors, MCI enlisted its law firm as a quasi-partner, tying it to MCI's success. Thus, MCI could do what few other AT&T competitors had ever been able to do: engage in a drawn-out legal battle with the telephone giant.

While MCI was keeping the pressure on Ma Bell in the courts, it also moved to branch out from the market of business-to-business communications services to residential long-distance service, attacking its nemesis at the heart of its operations. Despite McGowan's intense desire to expand, the move into residential service would be a huge risk. Americans had never before had a choice in long-distance service, and there was still a question as to whether they would even want one. To find out, MCI decided to conduct a test market in Denver.

Denver was chosen because of its appealing demographics and also because Mountain Bell was perceived as being more vulnerable than some of the other Bell operating companies. Although McGowan knew MCI could offer a good product, he believed his biggest problem would be convincing consumers that AT&T had worthy competition. And he decided it would be easier to gain a foothold in a city where residents weren't overly fond of Ma Bell's offspring to begin with. Taylor, who had been summoned back to Washington from Texas to head up the residential sales effort, turned to advertising and marketing as a means to this end.

Like any other service company, MCI had to deal with the fact that it did not have a physical product that a person could pick up and test or sample. It had no packaging that would draw attention to it. If consumers were to believe that MCI could offer them comparable service for less money than AT&T, it was advertising and marketing alone that would persuade them. Firm believers in the power of these tools, McGowan and Taylor spent considerable time working on ad copy themselves, but they realized that selling long-distance residential service to the 80 million U.S. households with telephones was vastly different than selling Execunet to sophisticated business executives. This market extended to every corner of the country and encompassed a much more diverse customer base. The opportunities for growth were extraordinary. Unlike previous generations who had generally reserved long-distance for emergencies or special occasions, a better-educated and increasingly mobile population often chose to pick up the phone and call just to keep in touch.

Deciding they needed some help, McGowan and Taylor enlisted Ally & Gargano, a New York advertising agency. Together with Tom Messner, a copywriter at Ally & Gargano at the time, they decided the best approach would be to go for the jugular and spoof one of AT&T's most successful ads. (Reverence had never been McGowan's long suit.)

Ma Bell's version featured a middle-aged couple sitting at their kitchen table. The wife is crying because her son has telephoned. Throughout the spot, the husband tries to elicit from his wife why their son Joey called, figuring something must be wrong. Finally, the wife manages to blurt out, "He called just to

MCI recreated an effective AT&T advertisement but added a new punchline to leave a lasting impression in the minds of viewers that MCI was the low-cost alternative to AT&T.

1. (MUSIC UNDER THROUGH-OUT) MAN: Have you been talking to our son long distance again? WOMAN: (NODS AND WHIMPERS)

2. MAN: Did he tell you how much he loves you? WOMAN: (NODS AND WHIMPERS)

3. MAN: Did he tell you how well he's doing in school? WOMAN: (NODS AND WHIMPERS AND CRIES)

4. MAN: All those things are wonderful.

5. What on earth are you crying for?

6. WOMAN: Have you seen our long distance bill?

7. ANNCR: (VO) If your long distance bills are too much, call MCI.

8. Sure, reach out and touch someone. Just do it for up to 30, 40, even 50% less.

MCI: Failure Is Not an Option

say he loves me." The ad was warm and fuzzy and almost everyone who saw it want to run to the phone and call Mom.

MCI's unforgettable takeoff on this Ma Bell tearjerker was almost identical. But at the end, when the husband asks, "What are you crying for then?" the blubbering wife responds, "Have you seen our long-distance phone bill?" The tag line was, "Reach out and touch someone—but do it for up to 30, 40, even 50 percent less."

Another spot designed to run at the same time featured two people making phone calls on a split screen with meters running at the bottom. On the AT&T side, the meter advances furiously, while on the MCI side, numbers roll by at a much slower pace. At the end of the commercial, the AT&T customer has racked up $6 in long-distance charges, but the MCI customer has spent only $3.07. In the voice-over, an announcer says, "You haven't been talking too much, you've just been paying too much."

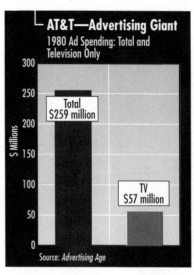

AT&T—Advertising Giant
1980 Ad Spending: Total and Television Only

Total $259 million

TV $57 million

Source: *Advertising Age*

Creating the ads, however, was only half the battle. Getting stations to run them was the other. AT&T fought the MCI ad campaign by lobbying the networks with an appeal to their sense of decency. It told them MCI was an unscrupulous company that was foisting a scam on the American public and that if they ran MCI's spots, they would be participants in this unsavory and potentially illegal activity.

Once again, AT&T was going to great lengths to thwart MCI, and because the Bell System was one of the biggest buyers of television advertising in the country, AT&T carried a lot of weight in the industry. With the big three networks looking for reasons to keep MCI off the air (and keep AT&T satisfied), Jerry Taylor had a tough time getting the ads to run. He had to meet with network executives face-to-face to convince them that MCI was a reputable organization and that the ads were merely humorous parodies.

One incident in particular causes Taylor to chuckle. "We had an ad where a grandmother was sitting on the porch in her rock-

AT&T's advertising budget was enormous. In 1980 it spent $259 million on advertising and $57 million on television ads alone. Its spending made it the ninth largest spender of all U.S. companies and an intimidating marketing competitor.

The People Behind the Images

The advertising minds who helped create MCI's image are a critical part of the company's success story. When MCI first became a client of Ally & Gargano, they worked with the talented creative team of Tom Messner, Ron Berger, and Barry Vetere. Together, they developed the AT&T spoofs that put MCI on the map.

Of course, it would be unfair to mention only their triumphs with MCI. While also at Ally & Gargano, the team was responsible for other legendary campaigns, including the "Time to Make the Donuts" campaign for Dunkin Donuts; the "When It Absolutely, Positively, Has to Be There" campaign for Federal Express; and the "Drive Safely" campaign for Volvo.

The three men left Ally & Gargano to form their own agency in 1986, joining forces with another ad man, Wally Carey, forming Messner Vetere Berger Carey. The name has changed over the years as partners came and went, including the 1988 addition of Bob Schmetterer. After Louise McNamee came on board in 1992, the firm took on its current full name of Messner Vetere Berger McNamee Schmetterer.

Messner has gone on to create more (legendary) ads including MCI's Anna and 1-800 campaigns, as well as the now-famous "Drive Safely" campaign for Volvo.

---

ing chair. When the postman brought her phone bill, she was supposed to flip entirely off the porch in a heap after she read it. And the networks said, 'This is ridiculous! How big could a bill be to make her flip over?' So they finally allowed her to fall over backwards in her rocking chair. That's how silly the discussions became. The fact was, the more ridiculous MCI made these spots, the more response we got."

When MCI finally launched the ads in March 1980 in the Denver test market, it received a thousand calls the first day. Totally unprepared for the deluge, its office ran out of order blanks within the first half-hour. Frantic operators began processing orders on the wall using any writing instrument they could find—including lipstick. MCI had no choice but to pull the ads and build up its staff to meet demand. "We were like a mosquito that had tapped into an artery," related Tim Price, president and chief operating officer of MCI. "We'd obviously hit a gusher."

True to its parent's tradition of doing everything possible to stop MCI, Mountain Bell disconnected MCI's phones for several hours that day, later claiming it was unintentional. But even that disruption could not stop the stampede, which AT&T may have unwittingly aided. On that same day, AT&T announced a 10.5 percent across-the-board increase in its long-distance rates.

Within a week MCI opted to speed up the residential-service launch in two other markets. By May 1980 it was so pleased with its residential response that it decided to sponsor a new, attention-grabbing promotion. To tweak AT&T during one of the year's busiest calling periods, MCI told callers nationwide: "In honor of Ma Bell, MCI gives its customers free long-distance calls on Mother's Day." One of the cleverest tag lines from that campaign belonged to Tom Messner: "Give Mom a priceless ring."

These ads marked the beginning of MCI's unholy advertising campaign against AT&T. They positioned the company as a wise-cracking upstart that was changing all the rules and became the foundation upon which MCI would build its future campaigns. Over the years those campaigns have sometimes been friendly, sometimes feisty, and sometimes even a little mean. But like those first flagship residential spots, they have always been unforgettable.

What also began in these early years was MCI's unyielding commitment to customer service. Because it had to fight so hard to get each and every customer, keeping customers happy with better, less expensive services was—and is—MCI's primary concern. "You've got a hell of a lot different attitude toward customers if you had to earn every one you get," explained Tim Price. In fact, of the ten largest carriers of international telephone traffic in 1996—MCI ranked number three—it was the only com-

In honor of Ma Bell, MCI gives its customers free long distance calls on Mother's Day.

1973–1980 ☎ Turning to the Courts

pany where all its customers consciously chose it. All the others started out as monopolies. (GTE, which was not a monopoly, was number eleven.)

MCI Gets Serious About Advertising and Marketing

Although McGowan and Taylor played a pivotal role in designing the residential advertising campaigns, it was Ally & Gargano that helped make them memorable.

MCI selected Ally & Gargano because the agency had put together a successful campaign for a fledgling package-delivery company called Federal Express. But equally impressive was Ally & Gargano's whole persona, one remarkably similar to MCI's—combative, spirited, freewheeling. "Truly, to know the personality of the agency, you would know that they would never get AT&T [as a client]," said Tom Messner. "Nor were they likely to get any major corporation that had any kind of pretension."

Ally & Gargano did comparative and negative advertising long before it was widely accepted. In fact, it was the first major agency to run an ad that named the competition. (In 1966, it attacked Avis, the nation's number two car-rental company, with the line: "For years, Avis has been telling you that Hertz is No. 1. Now we're going to tell you why.")

To develop MCI's campaign the agency organized focus groups to explore calling habits, pricing structures, and billing arrangements. Using information gleaned from the groups, the company was able to develop many successful spots. Among the more notable was an ad called "Big Business" that told residential customers how MCI had been saving its business customers money for years. MCI knew from the focus groups that although consumers disliked big businesses, they still respected them for what they considered smart and informed decision making. So the idea emerged to tell people how big names in business like

General Motors and American Express were using MCI to save millions of dollars on long-distance charges. By pointing out that the leaders in the business world had chosen MCI, residential customers would realize that MCI was the smart choice for them as well.

But the television spots that gave MCI real legitimacy with the public were those that used Hollywood celebrities such as Burt Lancaster. Even though MCI was going up against one of the most advertised companies in the world and was being outspent ten to one, MCI's ads had resonance. They gave the company a personality and indelibly etched MCI's image in the public consciousness. Moreover, they heralded the beginning of competition in an industry that previously had none.

American Express and spokesman Burt Lancaster helped lend legitimacy to MCI with the public.

Chapter 4

The Growth Years 1980–1984

> We'd run like mad, then we'd change direction.
> —BERT ROBERTS

The Growth Years 1980–1984

The decade of the 1980s ushered in a new era for MCI. No longer fearing imminent insolvency, it began to truly cast its net in distant waters. In the early 1980s Bill McGowan was talking about convergence while the rest of the industry was learning simply to spell the word Internet.

"Convergence of technologies," McGowan told his shareholders at the annual meeting in 1983, "that is, computers, information processing, and telecommunications, have all knit together to offer incredible new opportunities for new services, new products, and new industries—many of them not yet extant." Looking toward the full onset of the Information Age, McGowan added that he saw telecommunications as a crucial element in restructuring society.

From our vantage point in the latter part of the 1990s, McGowan's words seem eerily prescient. The Information Age is, indeed, witnessing a restructuring of society. We live in a time when e-mail travels around the world at the touch of a button, television news brings into our living rooms live images of a war being fought several thousand miles and multiple time zones away, and teenagers in Los Angeles listen to the same music and watch the same videos as teenagers in Bangkok. It is a convergence not only of technologies, but of cultures.

An Historic Antitrust Award

By the spring of 1980 MCI's initial antitrust case against AT&T was winding down. It had been among the longest, most expensive, and most complex civil proceedings ever tried before a jury. The discovery period alone took more than two years. Finally, on June 13, six years after MCI first filed its lawsuit, the jury came to a decision: AT&T had violated the Sherman Antitrust Act. MCI was awarded $600 million. Under antitrust statutes the amount was automatically tripled to $1.8 billion—the largest monetary award in American legal history at that time.

> Convergence of technologies, that is, computers, information processing, and telecommunications, have all knit together to offer incredible new opportunities for new services, new products, and new industries.
>
> – BILL MCGOWAN

$1.8 Billion AT&T Defeat

Los Angeles Times — Saturday, June 14, 1980 — Morning Final

Although MCI had grown into a substantial company by the time it decided to take on AT&T in the residential market in 1980, it was still dwarfed by its giant competitor. MCI's sales of $144 million seemed puny when compared to AT&T's awesome $51 billion in turnover.

"All the money, all the assets, all the people, and they could not win fair and square. They had to cheat," observed Larry Harris. "But they lost, and they lost big."

In Harris's opinion AT&T's unbridled arrogance throughout the proceedings probably drove the final nail in AT&T's coffin—an attitude obviously not lost on the jury. AT&T's people "thought they were a power greater than the government," he said. "It's hard to believe that kind of mind-set. But [they felt] if it wasn't in AT&T's interest, it wasn't in the U.S. interest."

This electrifying decision sent a powerful message to small businesses everywhere: No company, no matter how big, was sacrosanct. The giants could be beaten. The news took the business community by surprise. Very few believed MCI could win its lawsuit, and those who did never envisioned such a staggering monetary award. The stock market reacted with enthusiasm: MCI's stock shot up 21 percent the day after the announcement. Less surprising was AT&T's unwillingness to accept the jury's decision. While AT&T appealed, MCI didn't see a penny of the award. The case wasn't over yet.

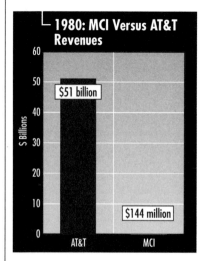

1980: MCI Versus AT&T Revenues — AT&T: $51 billion; MCI: $144 million

MCI Gets Its Passport

MCI wasn't about to wait for the elusive jury award. It had growing to do. For more than a decade, almost all of MCI's available funds were devoured by network construction, daily operations, and legal battles. But in the 1980s, with more revenue coming in from Execunet and from MCI's other private-line business offerings, and with new financing sources available in the

MCI: Failure Is Not an Option

MCI Profile

Judge Harold Greene

One of the most influential figures to emerge from the telecommunications tumult of the 1980s was the judge who presided over the government's antitrust case against AT&T—Judge Harold Greene. Although the Justice Department first charged AT&T with monopolistic practices in 1974, it was not until Greene's assignment in 1978 that the case began to progress.

He was widely admired for taking a firm hand over the proceedings—a critical skill in a case where evidence was so voluminous and complex. The proceedings took a decisive turn when he denied an AT&T motion to dismiss the case and commented that "the testimony demonstrate[d] that the Bell System had violated the antitrust laws in a number of ways over a lengthy period of time." Most importantly, he determined that the burden of proof was on AT&T. This commentary was critical in moving the case toward a settlement and finally toward the agreement announced on January 8, 1982.

Judge Greene remained responsible for interpreting the Modified Final Judgment (MFJ)—the decree that governed the Bell breakup—and was the sole arbiter on a number of other critical debates in the industry including whether to allow the regional Bell companies (RBOCs) to compete in long-distance. (He denied them.)

His influence was so absolute that it was questioned over the years. The *New York Times* described his role as "almost single-handedly deciding what telephone companies could and could not do." For those critical years after the AT&T breakup even the FCC and the Justice Department deferred to his decisions about the evolution of the regulatory environment.

In 1996 his official influence ended when he terminated the MFJ decree. His decision was in response to the passage of the Telecommunications Act of 1996, which opened up competition between local and long-distance telephone companies as well as cable and other communications providers.

1980–1984 ☎ The Growth Years

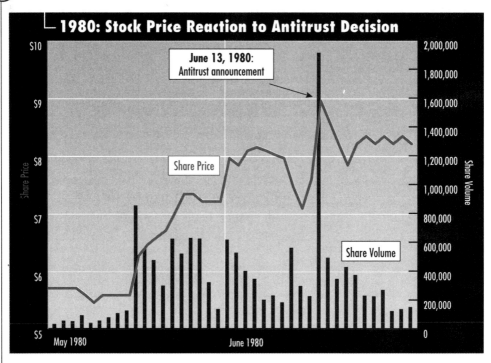

1980: MCI stock price reaction to the antitrust decision.

capital markets, the company began to search out ways to upgrade its service (which had always been sold on price rather than quality), as well as expand its business opportunities. While MCI was moving into domestic residential long-distance service in early 1980, it was also taking a close look at opportunities in the international market.

McGowan had long been aware of the potential of the $2 billion-a-year international market and considered it to be an integral part of MCI's expansion strategy. Its rapid 20 percent annual growth rate, which far surpassed the 10 to 15 percent annual growth rate in the domestic long-distance market at the time, made it impossible to overlook. AT&T, however, also had a monopoly on U.S. telephone traffic overseas. McGowan was willing to take on AT&T globally, but Orville Wright and other key managers convinced him that the expense would be prohibitive.

They chose instead to find their international foothold in data transmission as an international record carrier (IRC). The IRCs, through agreements with foreign postal, telegraph, and telephone ministries, controlled the messaging business, which primarily consisted of telex service. Telex is short for teletypewriter exchange service and is like telephone service except that typed messages are transmitted between teletypewriters at either end of the line. The sender types a message, dials the receiver's number, and forwards the message, which is then printed at the other end. Although the telex has been largely replaced by the fax machine, it is still used in less developed parts of the world.

What made the idea particularly appealing was a recent FCC ruling that permitted AT&T to enter the messaging and data market and in turn allowed the IRCs to provide voice service. Because of these regulatory changes, MCI could simply sidestep its way toward a full frontal assault on AT&T's international voice monopoly by acquiring an IRC.

In December 1981 MCI struck a deal with Xerox Corporation to purchase its Western Union International (WUI) subsidiary for $185 million. Ironically, WUI had negotiated with Goeken to buy a 50 percent interest in MCI back in 1968, before McGowan came on board. But it eventually backed out, believing MCI would never get a license from the FCC.

From left to right are Bill McGowan; Nathan Kantor, who was in charge of the MCI team that investigated international carriers (later president of MCI International); and Edward Gallagher, president of WUI.

Although WUI provided the springboard MCI needed to enter the international market, the marriage took considerable time to mature. The two companies were vastly different. Founded in 1963, WUI was part of the traditional, regulated world of telecommunications—much like AT&T—and enjoyed a guaranteed rate of return. WUI adroitly used regulation for protection while MCI bristled at regulation. In addition WUI was unionized and overstaffed.

Merely changing WUI's name to MCI International, did not bring about an easy transition. There were many assimilation problems. WUI's union workers went on strike, partly in protest of MCI's layoffs of WUI personnel.

Cutting employees was a job Bill McGowan never liked. There were times, of course, in MCI's history when the company had had no other choice but to reduce staff in order to keep alive. The early years had been a financial roller-coaster ride.

Xerox: The Document Company

From its 1906 beginnings as the Haloid Company, a maker of photographic paper, there emerged a company whose copying process—named xerography from the ancient Greek words for dry and writing—went on to revolutionize both the business world and the world at large.

Few may remember what the state of the art looked like before the first simplified office copier, the Xerox 914, was introduced in 1959. Mimeograph, thermal paper, and damp copy couldn't hold a candle to the process whose shortened name became synonymous with both the verb and noun forms of the word photocopy. By the time MCI agreed to purchase WUI in 1981, Xerox Corporation was diversified as a maker of printers, plotters, and disk drives as well as the owner of three publishing firms and one computer company. Many of these interests were subsequently sold or discontinued, and Xerox moved in 1990 to refocus on its core document processing lines.

When money ran out at the end of 1973 and network construction ground to a halt, workers were laid off. Then again between 1974 and 1976 when the legality of Execunet was in question MCI had been forced on three occasions to lay off employees. Often, though, people were brought back once the cash crisis had eased. In June 1976, for example, fifty employees were let go, only to be called back in July.

In this case, however, WUI was grossly overstaffed, and there was virtually no chance these employees would be rehired. The strike, the first in MCI's history, lasted three months before the union ratified a three-year contract, but it took several years before WUI would integrate and approach the level of efficiency and empowerment that characterized MCI's workforce.

MCI Reaches the High Ground

MCI's eventual profitability and growth can be attributed to many things. "We had drive, we had determination, and a lot of Irish luck," observed Jerry Taylor. Good fortune and Taylor's humility aside, MCI did a lot of things right, such as hiring good people, making key corporate acquisitions, providing good customer service, pushing regulatory changes, pursuing lawsuits, and employing innovative advertising and marketing programs. MCI could have accomplished nothing, however, if it were not, first and foremost, successful at obtaining financing. "The only thing you can never recover from is running out of cash," remarked Bill Conway, who joined MCI in 1981 as vice president and treasurer, having previously been MCI's loan officer at First National Bank of Chicago.

Nonetheless, McGowan was never keen on selling equity. The day of MCI's 1975 offering, when the company was forced

MCI International

The WUI acquisition was most significant because it marked the beginning of MCI's goal of becoming a completely global company. Milestones throughout the years include:

1982   MCI purchases WUI from Xerox, giving it entree into the international arena.
1988   MCI acquires RCA Global Communications, Inc. (whose principal businesses were worldwide telex, facsimile, and high-speed data transmission) from GE.
1992   MCI enters Canadian market through alliance with Stentor, planning the first integrated network between Canada and the U.S.
1994   MCI forms alliance with Grupo Financiero Banamex-Accival (Banamex) to provide long distance service in Mexico. The group begins construction of fiber optic network linking all major Mexican cities in preparation for opening of market to competition in August 1996.
1994   MCI and British Telecom announce alliance to attack global markets. BT invests $4.3 billion for 20 percent of MCI. Their joint venture has a goal of providing a full range of services to multinationals.

to surrender 39 percent of its equity at depressed levels, McGowan later described as the worst day of his life. He vowed never again to put the company in such a position.

Needing cash to bankroll MCI's expansion, McGowan and English had turned to different forms of financing. Between the end of 1978 and October 1980, they raised almost $150 million through three offerings of convertible preferred stock. Then in the summer of 1980, another new financing vehicle appeared, subordinated debt, which would provide the lion's share of MCI's capital needs in subsequent years.

Subordinated debt can be used to finance a growth company like MCI without diluting equity—a very attractive feature as far as McGowan was concerned. But since it is secured only by the issuer's promise to pay, it usually carries a somewhat higher interest rate than secured debt in order to compensate investors for the increased risk.

MCI's first offering of subordinated debentures, twenty-year notes paying 15 percent interest, was well received by the financial world and netted $50 million for the company. Some $45 million of this amount went to pay down bank loans. Equally significant for MCI's future, however, was the discovery and successful offering of this financial vehicle, which provided a way of raising more money on longer and more flexible terms.

The convertible preferred offering that followed in October brought in another $51 million and allowed MCI to wipe out its bank debt—no small matter because at the time the prime rate had reached a staggering 19.5 percent, which in turn had pushed MCI's interest payments to distressingly high levels. The offer-

MCI Profile

William E. Conway

Bill Conway first worked with MCI in 1975 when he was a junior loan officer at the First National Bank of Chicago, which was the company's bank. He eventually rose to become the youngest vice president in the bank's history.

Conway joined MCI Communications Corporation in 1981 as vice president and treasurer. In July 1982 he was named senior vice president of finance for its long-distance subsidiary, MCI Telecommunications.

He later succeeded Wayne English as chief financial officer for the entire company. With MCI's financial future firmly established, Conway left MCI in 1987 to become managing director for the Carlyle Group, a merchant banking company in Washington, D.C.

ing also freed up the company's capitalization by allowing it to phase out equipment leases. MCI's strategy was to eliminate this senior secured debt with as little dilution of equity as possible. The company could then begin the process of building its capital structure from the bottom up.

Hot Issues—MCI Explores New Financing Strategies

MCI had come a long way in the two years since Wayne English launched his plan to restructure the company's finances. It had raised over $200 million from four public offerings, and had transformed itself from a company with a $30 million negative net worth in 1978 to one with $148 million in positive equity in 1981. With its new status as a profitable company and taxpayer, MCI was now able to take advantage of the debt markets while continuing to build its capital structure. The company began to explore various debt strategies, particularly ones that would give it maximum flexibility in financing its expansive growth.

In the early 1980s MCI debt was still ranked as non-investment grade or junk, meaning that the major credit-rating agencies considered the company to be a risky investment. The low rating in turn meant that MCI had to pay investors more interest to compensate them for taking a bigger risk.

High-yield, or junk, bonds were nothing new, of course. They had been used for many years by companies that did not qualify for investment grade status. These companies generally fell into two categories: small, usually young companies with little or no credit history, and fallen angels—larger, better established companies that had fallen on hard times. High-yield bonds had long played a critical financing role in the growth of such companies, but around 1977 they began to take on a new importance.

The growth of the high-yield market was more than another Wall Street financing innovation—it was a natural development of the economic lessons of the 1970s. For investors the credit crunch of 1974 had illustrated that a company's creditworthiness wasn't determined by how long it had been in business or by its size. Investors began to focus more intently on the individuals managing businesses, looking beyond the historical in order

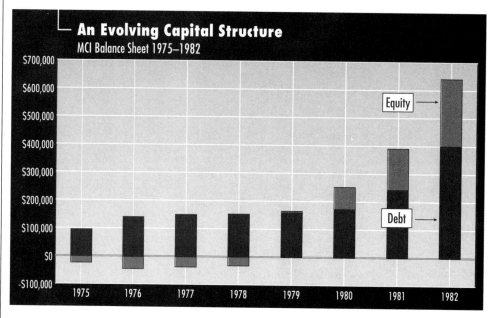

As MCI built its equity base, it commanded more respect in the financial markets.

to discover those companies with the people and vision needed to prosper in the years ahead.

The credit crunch had changed the way companies viewed financing as well. After their experience in 1974 during which they had been essentially cut off from the conventional banking credit sources, many companies were anxious to develop other sources of capital. Like people who had lived through the Great Depression to become lifelong savers, those entrepreneurs and managers could not forget the financial disruption that had led to canceled plans and lost opportunities. These memories caused many of them to raise long-term fixed-rate capital as means of ensuring financial freedom. Additional capital was viewed as an insurance policy, one that could guarantee a company's ability to implement its plans for the future, independent of banks and other financial institutions.

As this period of growth dawned in the late 1970s these two groups—companies (such as MCI) who needed capital to grow and the investors who needed attractive investment opportunities—came together in the nation's public capital markets. Non-investment grade companies who sought long-term fixed-rate capital found investors willing to purchase a diverse group of securities including common and preferred stock, convertible securities, straight debt, and commercial paper. In 1980 the high-yield segment suddenly doubled from its level a year earlier, raising $1.4 billion for new companies. That year marked the start of a major expansion. In 1995 the junk market was responsible for $28 billion in new financing.

The economy was also primed for growth. Despite record-high interest rates in 1980 and 1981, the capital markets—unlike the situation that had occurred in 1974—did not shut down. Money was available to corporate borrowers; it was just very expensive. In this environment a multitude of financing options were being created, and Drexel Burnham Lambert was in the forefront just as high-yield bond issuance started to take off.

In the first six months of 1981 fifty-nine new high-yield issues came to market with proceeds totaling more than $4.1 billion. That was seven times the proceeds from such offerings during the first half a year earlier. Also, the spread between U.S. government and non-investment-grade bonds (rated Double-B) was relatively narrow at about 300 basis points. That meant that non-investment-grade companies could take advantage of the

In 1980 there were only 45 high-yield underwritings, raising about $1.4 billion. By 1995 the number of underwritings had risen to 154, and the total dollar amount raised was $28 billion—a twentyfold increase over fifteen years.

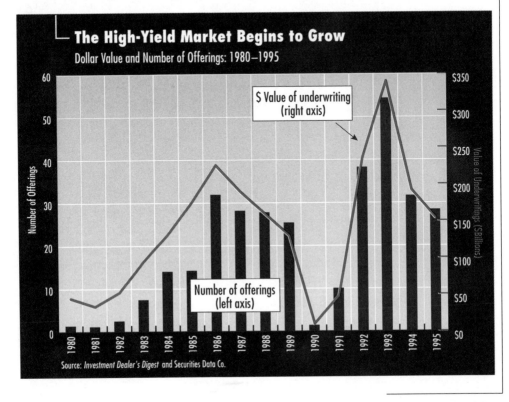

The High-Yield Market Begins to Grow
Dollar Value and Number of Offerings: 1980–1995

Source: *Investment Dealer's Digest* and Securities Data Co.

1980–1984 ☏ The Growth Years

New Ideas Transform the Markets

Before a new breed of financiers transformed the capital markets, non-investment-grade companies had very few financing options. They could raise capital either by borrowing from the banks, which usually meant short-term floating-rate loans, or they could privately place senior or secured debt offerings with insurance companies. Both types of borrowings were weighted down with restrictive covenants that inhibited the borrowing company's growth. Additionally, private placements often included equity kickers, anathema to entrepreneurs because they gave away so much of a company's equity upside.

bond market's long-term, fixed-rate financing opportunities at only slightly higher interest rates than they would have to pay on other types of financing.

In the 1970s and 1980s no company was more successful at marketing securities for high-yield companies than was Drexel Burnham Lambert. MCI's management team was well aware of this expertise—and of Michael Milken, a bond trader and investment banker who specialized in high-opportunity, high-growth investments. By the end of the 1980s Drexel would have raised more than $200 billion through sales of both debt and equity securities for companies previously considered out of the mainstream.

The similarities between McGowan's team at MCI and Milken's team at Drexel were striking. Every day on the trading desk, Drexel employees believed they were accomplishing something that was innately worthwhile—going to bat for the little guy. The High-Yield and Convertible Bond Department exuded a vitality and creative spirit that was readily apparent to its clients. And like McGowan, Milken was a visionary leader who was constantly searching for new avenues of growth. Because of this marked tendency to go against convention, Milken accepted challenges that others shied away from and financed industries that have changed the way the world does business. It was a heady time for all who were fortunate enough to play a part.

Milken had begun to build his reputation in the investment community in the 1970s early in his Drexel career when he identified undervalued securities, equity as well as convertible and straight bonds. Many of the bonds were trading at deep discounts from twenty cents to seventy cents on the dollar.

Milken convinced Drexel to develop a credit research group that would follow corporate debt issues. (At the time, most firms only did equity research.) Subsequently, he recommended many of these securities as buys. When they appreciated in value rather than failed, as most experts had predicted, the investment community took notice. Here was a man who could spot opportunity where others saw only trouble.

By 1977 as part of the firm's evolution in the marketplace, Drexel turned this expertise to financing new companies and refinancing companies in distress. And just as MCI had its eye on Milken, so too Milken had his eye on MCI. He had followed the company for many years and was convinced he could help management create value. But there was one big obstacle: I. W. (Tubby) Burnham, Drexel's chairman. Burnham, who was on the board of directors at Continental Telephone, discouraged the firm from financing a competitor. Milken knew, however, that MCI's determined management was going to get financing with or without Drexel. Like so many in the business world, he was taken with MCI and wanted the chance to participate in its future.

Milken has an uncanny ability to analyze quickly a company's financial condition, its management's vision and ability, and to evaluate its products and services. In addition Milken recognizes the inextricable link between business and society. In MCI's case, he knew that its ability to understand the changing nature of society's needs and desires would be recognized in the marketplace. He also knew that if MCI was going to compete effectively against AT&T, it would need plenty of capital.

Finally, in the spring of 1981 with a lot of prodding from Milken and Drexel's corporate finance department, Burnham relented. Wayne English was on a plane to Los Angeles within a week. McGowan and Conway soon followed.

During the due diligence process between the two companies, MCI's management team spent considerable time describing their hopes for the future and their strategies for expanding their business. Milken was full of ideas and quickly came up with a plan for financing MCI's future, beginning what would be a

 MCI Profile

Michael Milken

Although the general facts of Milken's career are widely known, the extent of his contribution to the telecommunications industry in particular is less well recognized. His involvement has been lengthy and far-reaching and has made him one of the key players in the history of this tumultuous industry.

Key to his success in the information and technological industries was his natural love of their basic businesses. Indeed, much of his success as a bond trader came from the changes he made in the systems processes at Drexel Burnham Lambert. (He had specialized in information systems at Wharton.) He began by computerizing the securities-delivery system (saving Drexel $500,000 in interest charges), and later by overseeing the creation of an advanced trading system for his team. The system set a new standard in the industry and was critical to Drexel's dominance in the high-yield (junk) market. Traders in Milken's office had the unique capability to quickly call up a customer's trading history, investment limits, and other key trading information. The system was a major success because it could match buyers and sellers more effectively than had ever been done before in the chaotic high-yield world.

His affinity for information processes extended into his investment banking interests. His financing deals provided critical capital to scores of (then) small start-up companies in the communications industry. The names of these companies are familiar to us today as global leaders in many different areas of information technology: Aside from MCI, he provided start-up capital to McCaw Cellular, Viacom, TCI, Time Warner, Turner, Cablevision Systems, and News Corp., to name a few.

Milken's success was based on his recognition of the importance of information. He studied the economy and listened carefully to the stories of the companies who came to him for financing. He also saw that the global economy was shifting to one in which information technology, rather than manufacturing superiority, would be the key to economic advantage.

long and productive relationship with McGowan and MCI. The ability to provide companies with more than just financial advice is the quality that made Milken stand out in the investment banking community. He worked with companies on all aspects of their strategic planning—from financing to marketing—with the goal of creating value for both the company and its investors.

Duly impressed with Drexel's abilities, MCI invited the firm to participate in its next public offering even though they knew that allowing Drexel to co-manage the deal could very well upset Shearson, which had been MCI's underwriter since 1979. (Most of the rivalry between banking firms occurs in the corporate finance departments. As it turned out, Drexel actually had a good working relationship with Shearson's high-yield sales and research departments, and the two firms were able to work together to raise capital for MCI without too much friction.) But McGowan, Conway, and English knew they would need to defy all obstacles and had the foresight to recognize what Milken would bring to the company. They knew he was going to help them. What they probably didn't realize was exactly how much.

MCI Becomes One of the Hottest Properties in the Bond Market

MCI had its strategy firmly in place. In April 1981 $125 million was raised in the first co-managed offering, selling twenty-year subordinated debentures. With its net proceeds of more than $100 million, MCI planned to pay off $15 million of revolving credit debt while using the balance for network construction. Less than a week later, it redeemed all of its convertible preferred stock from the 1979 offering.

Another offering followed in August, this time consisting of $100 million of twenty-year convertible subordinated debentures with all proceeds earmarked for network construction and new equipment. Seven weeks later, MCI redeemed the convertible preferred stock it had issued just the year before.

One of Milken's strengths is his ability to help a company manage the left side of the balance sheet—its capitalization. And although he was not a particular fan of convertible bonds, he did believe they had their time and place in a well-thought-out financial strategy. Building a capital structure requires a long-term plan, and Milken and McGowan were following the plan they had devised for MCI: to build a foundation of subordinated offerings without causing shareholder dilution.

For MCI, whose stock price was on the rise, Milken recommended convertible bonds that could be exchanged into stock in a short period of time. In the August 1981 offering, for example, the bond was convertible at about $25—almost 20 percent higher than the stock's trading price at the time. After the stock price took off, the bonds were called in February 1983 and converted into common stock, increasing MCI's equity account by $100 million.

In May 1982 MCI proposed yet another issue—$100 million of twenty-five-year convertible subordinated debentures. But considering the favorable economic conditions and the recep-

Debt Versus Equity
April 1981: Financing Decision

	March 1979	March 1980	March 1981	
Long-Term Debt	$159,035	$172,852	$242,707	
Preferred Stock	$839	$742	$1,087	
Total Shareholders Equity	$5,754	$78,847	$148,047	
Leverage: LT Debt/Total Cap	**97%**	**69%**	**62%**	- Leverage coming down
Share Price (unadjusted)	$3-2/3	$6	$13-1/4	- Stock price rising...
Market Capitalization	$140,256	$427,975	$1,605,013	- along with market capitalization

The subordinated debentures issued in April of 1981 marked MCI's first collaboration with Drexel Brunham Lambert. At Drexel's recommendation, MCI chose to issue subordinated bonds for a number of reasons. First of all, it had successfully deleveraged during the previous two years and felt it had room to take on more debt. Its stock price was also on the rise, giving the managers reason to believe that it would have ample opportunity to issue equity in the future and that this equity would come in at higher prices. This assessment turned out to be particularly prescient as its stock price staged its first major rally throughout 1981–1982, leading to MCI's first stock split.

MCI—Money Coming In

Three Years of Fast Financing

April 1981	$125	14-1/8 percent subordinated debentures
August 1981	$100	10-1/4 percent convertible subordinated debentures
May 1982	$250	10 percent convertible subordianted debentures
September 1982	$250	12-7/8 percent subordinated debentures
March 1983	$400	7-3/4 percent convertible subordinated debentures

tiveness of institutional investors, Milken urged the company to increase the amount to $250 million. "The best time to finance is not when you need the capital," he always maintained, "but when the market is most receptive to providing capital—not when your back is against the wall." MCI followed this advice.

Institutional investors snapped up the issue. MCI used the proceeds from the offering to pay for its acquisition of Western Union International and for further network construction and new equipment.

Part of Drexel's job was to get institutional investors interested in MCI's future. Working together with the company, the firm helped build long-term relationships. Through its credit research and the support it provided at road shows where the company management actively marketed its financing issues to the public, Drexel managed to attract a following that stayed with MCI through good times and bad.

This work paid off for MCI. Investor interest in the company was growing along with what Bill Conway described as its "voracious" appetite for capital (it was spending $50 million a month), so MCI continued to file new offerings. The next one was for $100 million of subordinated debentures in September 1982.

"We'd finish one offering and then start talking about the next one," English recalled. The strategy clearly worked: In one year of issuing debt securities, MCI raised more than twice as much capital—$900 million—as it had raised in the previous ten years combined. At the end of fiscal 1982 MCI had $144 million in cash, an equity market value of $1.6 billion, and stockholder equity of $241 million.

> The best time to finance is not when you need the capital, but when the market is most receptive to providing capital—not when your back is against the wall.
>
> – MICHAEL MILKEN

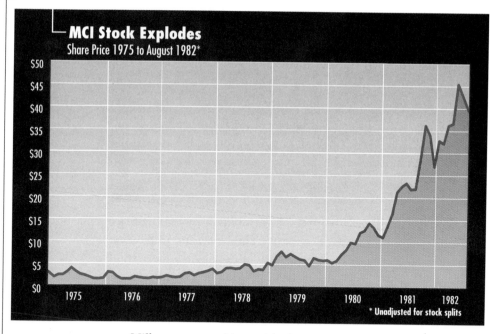

MCI share price exploded in 1981, leading to the company's first two-to-one stock split later that year.

Milken was working with the company and McGowan to attain financial freedom. If MCI were to survive and compete effectively against its deep-pocketed nemesis and other competitors in this capital-intensive industry, it needed to build for the future. With each successful offering, MCI was closer to making that dream a reality. But the company was not out of the woods yet.

The Need For More Capital

Complicating MCI's quest for financial freedom were substantial technological limitations. Execunet's booming growth was overwhelming the capacity of the microwave system just as environmentalists stepped up opposition to its expansion. A network based on fiber optic cable appeared to be the answer to MCI's problems, but even more money would be needed to build it.

A question remained as to how such a relatively small company as MCI could acquire the enormous amount of cash to accomplish its goals. Despite MCI's meteoric ascent (its equity market value had increased sevenfold since 1979), it was still struggling against AT&T, its primary competitor and the largest private enterprise in the world with $66 billion in revenues. With only $506 million in revenues, MCI was less than one one-hundreth the size of AT&T.

Despite its comparatively small size, the market knew that MCI was a formidable contender. Drexel's efforts to capitalize MCI and build its equity base were rewarded in March 1983 when Moody's Investors Service upgraded its rating of MCI's debt to Baa3—investment-grade status. Although Standard and Poor's, the other dominant credit-rating agency, still rated MCI as non-investment grade, the import of Moody's seal of approval was tremendous. Just a few years earlier, most industry experts had viewed MCI as nothing more than a fly on an elephant's back. Now it was honored with a rating that only 5 percent of all American businesses attain. (The debt of nearly 95 percent of all businesses even today is regarded as junk—below investment grade.) It was, by any measure, an important achievement.

This elite ranking would also make a big difference in the way MCI priced its future bond offerings, enabling it to lower the amount of interest it would have to pay out on its debt issues. Furthermore, as a result of the Moody's upgrade, a new class of investors would be drawn to MCI as a long-term investment.

Empowered by its new status and driven by the constant need for capital, Milken and McGowan determined that the time was right for MCI to raise additional money. The company decided to sell $200 million of convertible bonds. The Drexel High-Yield and Convertible Bond Department once again was gratified by the results: Investor demand was so strong that the company doubled the size of the offering and easily sold $400 million of twenty-year convertible bonds.

With this sound financial strategy, MCI was well on its way to building its capital structure. In short order, the company would be able to force conversion of its convertible bonds into common stock, thus in one stroke eliminating some of its debt and building its equity base—and garnering a higher price for the stock than it could have gotten at the time the bonds were issued.

Why Convertibles?

MCI liked convertibles for a number of reasons:

- They fit into its strategy of building its capital base—once the bonds were converted they would become equity on the balance sheet.
- In this manner, MCI also received a higher price for its stock as it effectively delayed issuance until its share price had risen.
- The May 1982 convertible bonds were a textbook example:

① May 1982—$250 Convertible Bonds Issued at 10 percent

- Share price at issuance: $4.65
- Convertible at $5.625 (21 percent premium)
- MCI may force conversion only if share price 150 percent of conversion (i.e., $8.44) for 30 consecutive days

② June 1982—Share Price* Takes Off

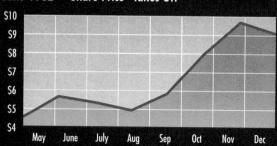

③ December 1982—MCI Calls for Conversion

Benefits to MCI

Debt turns into equity
Debt $246
↓
Equity $246

Higher price for stock
$5.625 conversion price versus
$4.65 at bond issue
= **20% higher price**

Benefits to Investor

Profit on share price rises
Convert bonds into stock at $5.625 versus
$9 price on market on conversion   = **60% automatic profit**

* Adjusted for stock splits

The association with Milken proved to be a defining event for MCI. "We wouldn't be in business today without him," reflected Jerry Taylor. Milken used his own expertise and reputation for value to create demand and institutional support for MCI and its securities in the capital markets. Moreover, Milken provided the guidance to strengthen the company's capital structure by creating the right blend of financing to help MCI get through the tough years that lay ahead.

MCI Switches Technological Gears

With its astounding success in the capital markets, MCI now had the ability to start out on its path of self-evolution. MCI ordered more than 150,000 miles of fiber optic cable. These orders, the largest ever placed, ushered in a new age for the telecommunications industry.

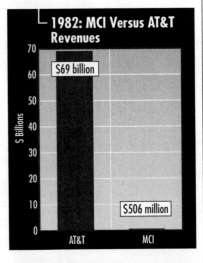

The idea of using light to carry sound or data was not new, but the technique hadn't been perfected until 1970 when Corning Glass developed a glass cable. Fiber optic cable proved vastly superior to wire or microwave transmission. It transmitted high-quality signals that were not affected by weather conditions or geography and that did not interfere with other frequencies.

MCI was the first company to take a serious look at the practical application of fiber optics after it realized that Execunet's rapid growth would soon exhaust its microwave transmission capacity. Microwave construction had become increasingly difficult and expensive, and activists had begun to protest the aesthetic and environmental damage from microwave towers, thus making zoning approvals harder to obtain. (AT&T, which was still using a traditional copper-wire system, had few such problems.) MCI's chief engineer, Tom Leming, who oversaw the building of the original system, was looking for al-

MCI pioneered single-mode fiber optic cable. Each strand of glass could carry 6,000 voices at the same time.

Pulling fiber cable through the tunnel under the Hudson River proved less of a challenge than crossing an Arkansas swamp, which nearly swallowed the cable tractor.

ternatives. He had followed fiber optics development and believed that single-mode fiber cable offered the best solution. Less expensive than the multimode fiber invented by Bell Labs and favored by its parent AT&T, single-mode cable carried the light impulses longer distances without the need for additional repeater devices, and it also provided a higher transmission capacity.

As an added advantage, fiber optics would go a long way toward addressing MCI's quality issues. Although consumers chose MCI for price, the perception that AT&T had better-quality connections was something that MCI wanted to change.

For all its advantages, purchasing the fiber optic cable still represented an expensive and intimidating change of direction. However, after "60 Minutes" aired a show about the dangers of microwave radiation, the move to fiber optics seemed all but inescapable.

Where To Lay The Cable?

The problem of where to lay the cable had been perplexing at first. Leming had considered interstate highways, but federal regulations ruled out that option, and state and local roads were not viable because the delicate cable would be exposed to frequent road construction. Leming came up with an ingenious solution, one that quite literally encompassed the entire country. Railroads, he determined, were the answer.

After lengthy negotiations, his staff leased a right-of-way from Amtrak, the passenger train company, between New York and Washington, D.C. MCI also forged an agreement with railroad carrier CSX Corporation in mid-1983, covering its routes between Chicago, Philadelphia, Detroit, Pittsburgh, Baltimore, and cities in the Southeast. This accord gave MCI access to rights-of-way east of the Mississippi as well. It would take several years to complete the railroad deals and the installations necessary to make fiber optics a reality, but the move away from microwave was definite.

This embrace of fiber optics epitomized MCI's willingness and ability to reinvent itself when necessary. Although microwave transmission was the technological foundation on which the company was based, MCI faced the reality that this system was becoming obsolete and adapted quickly. Equally significant was its ability to take the long view even when mired in immediate problems such as the AT&T antitrust proceedings.

Converting to a fiber optic network was inordinately time-intensive and costly: Capital expenditures for expanding and upgrading the network ran at around $20 million a week, or about $1 billion a year, in the mid-1980s! Committing to fiber optics was also technologically risky, considering the lightning speed of technological evolution. Yet, MCI knew that long-term success depended on establishing a state-of-the-art nationwide network. With its usual aplomb, it set out to stay at the leading edge of technology and in so doing led the way for the entire telecommunications industry.

A Satellite Sidetrack

While American astronauts were piloting reusable space shuttles in what *National Geographic* magazine called "the second space age," MCI was negotiating to get its own space adventure under way. In February 1983 the company announced its first acquisition of satellite capacity—the largest in telecommunications history.

Shortly after ordering its 152,000 miles of fiber optic cable, MCI agreed to purchase twenty-four satellite transponders, or radio relay devices, from Hughes Aircraft, a California-based aerospace company. The purchase sparked an intense debate among MCI executives over whether satellite technology or fiber optics should become the heart of MCI's transmission system.

McGowan, who had long advocated the application of space-age technology, believed that satellites offered a level of flexibility over landlocked systems. But he also believed that the two technologies could be used effectively together. Tom Leming and Orville Wright disagreed. They were concerned that satellite

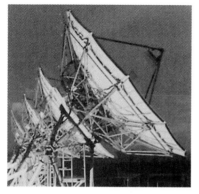

transmission would involve too much delay and echo for voice use, which customers would simply not tolerate. They favored putting the money into more fiber optic cable.

McGowan considered their arguments carefully, but eventually decided to invest in the satellite technology. Plans for the construction of a satellite network at a cost of $200 million to $300 million were begun. Still, Wright and Leming's position ultimately won out as single-mode fiber optic cable became the standard for long-distance carriers—thanks to MCI's pioneering efforts. The ability of fiber optics to carry high volumes of traffic quickly, efficiently, and with superior quality gave it the edge over satellite technology at that time.

But even though satellite technology did not really fit into MCI's plans in the early 1980s, McGowan's interest in it perfectly exemplified his innovative thinking. Once again, he was far ahead of the rest of the industry. As the new millennium approaches, the telecommunications industry is looking to direct broadcast satellite systems for transmission of digital-quality television programming, music, voice, data, and even software.

The willingness to try new things continues to give MCI an edge over its competitors. What might be criticized in some companies as erratic behavior is applauded at MCI as strength through flexibility. "If we don't make the right decision," remarked Angela Dunlap, MCI's executive vice president of corporate communications, "we change it and move on."

MCI did change its satellite strategy because of the persistence of Leming and Wright. But it was never easy to maintain an opposing position in the face of McGowan's intelligence. It was not that he was particularly difficult or that he instilled fear; in fact, McGowan loved hearing new ideas. It was just that he was so rarely wrong. One manager describes the phenomenon as the Ted Williams effect. Just as an umpire silently questioned his strike call if the Boston Red Sox batting champ failed to swing at a pitch, MCI managers frequently questioned the soundness of their ideas if McGowan didn't swing at them.

The satellite diversion didn't stall MCI. The company was accustomed to zigzagging its way toward growth. "We'd run like mad," recalled Bert Roberts, "then we'd change direction."

> We'd run like mad, then we'd change direction.
>
> – BERT ROBERTS

A Steady Stream of New Products and New Alliances

While MCI's management debated these shifts in technology, the company's marketing machine continued to break new ground as well. New products were created to tap into completely new markets and customer needs. MCI's first WATS service, for example, was designed for customers spending a minimum of $1,500 per month. It matched AT&T's service but at a 20 percent discount. This period also marked the introduction of MCI's no-fee telephone credit card, which saved customers 40 percent over AT&T's calling-card charges. And MCI also introduced its first public Charge and Save telephones, which could be accessed using MCI's charge card, at Washington, D.C.'s National Airport.

Besides rolling out innovative offerings at home, MCI stepped up its efforts to capture international business. Earlier plans had focused on launching international service via the message and data arena, but the telex market was shrinking (as fax machines became popular) and MCI continued to have union trouble with the former WUI (now MCI International). So the decision was made to attack the voice market head-on, and Canada was selected as the most logical point of entry.

Canada was the single largest segment of the $4 billion foreign market with an estimated $670 million in potential annual revenue. It was an English-speaking country (with the exception of Quebec), and with nearly 90 percent of its population living within a hundred miles of the U.S. border, Canada was a natural spillover audience for MCI's television campaigns. MCI's early service had severe limitations in that it was one-way (from the United States to Canada) and only went to Montreal, Toronto, and Calgary, but it represented an important jumping-off point.

Australia, the largest English-speaking country in the Pacific Basin, and Europe were next on the list. MCI reached an agreement with Australia in late 1983, but the European market proved tougher to crack. Although Belgium was won over by June 1984, Europe's other state-owned postal, telegraph, and telephone ministries (PTTs) remained firmly committed to keeping their agreements with AT&T.

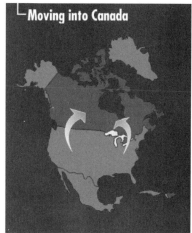

In 1983 MCI entered the Canadian market.

To break the stalemate, MCI needed a major European partner, and it considered British Telecom (BT) to be the best candidate. Beginning in the early 1980s, MCI appealed to BT by explaining how competition could lower rates, stimulate traffic, and increase the revenue collected by the British government. MCI also argued that country-specific advertising would produce "double stimulation" by increasing both the number and length of calls.

MCI's case for competition based on market stimulation and revenue enhancement was as heretical to the PTTs as Goeken's private-line microwave idea had been to AT&T twenty years earlier. For three-quarters of a century, international carriers had not advertised. Why should they? There was no competition. Now, MCI was pushing to change all that, and BT was wary. Although recently privatized by Margaret Thatcher's Conservative Party government, BT continued to think like the state-owned monopoly it had been for so many years.

Skeptical of MCI's arguments and worried about disrupting its relationship with AT&T, BT continued to delay reaching an agreement. Only when Thatcher's government promoted BT competitor Mercury Communications' bid for increased domestic competition did BT succumb to MCI's persuasions. Fearful of being left behind, BT eventually signed a contract with MCI in October 1984 that provided for direct-dial service beginning in 1985. The BT agreement was the wedge MCI needed to expand its international network to the rest of Europe.

(Like AT&T, BT was being forced to face competition—and soon it will face even more. The European Union has authorized full-scale competition in 1998. Moreover, in a surprise move at the beginning of 1996, it began allowing "alternative networks," such as utilities, to offer telecom services.)

Financial Freedom: The Ultimate Offering

Although Drexel had been successful at helping MCI attract financing, it still had a long way to go toward fulfilling MCI's long-term capital needs. Then, in the summer of 1983, the opportunity arose to give MCI true financial freedom.

During that summer Wayne English convinced McGowan and MCI's board to go ahead with a $500 million issue of bonds with attached warrants. But Milken told McGowan that Drexel was prepared to raise more than a half-billion dollars—much more!

The innovative offering Milken designed, debt with an equity kicker, would effectively create value for both MCI and investors. The issue was structured to appeal to four different types of security buyers, thus widening the potential market. The warrant component appealed to pure equity players. The unit as a whole was a synthetic convertible, which appealed to convertible security buyers. And the bond portion appealed to two kinds of straight debt buyers—those who could buy high yield and those who preferred high grade investments. (Remember, Moody's had bumped MCI up to investment-grade, but Standard & Poor's still rated its debt as junk.) Most importantly, the security could be divided into three separate instruments: bonds, warrants, and bond-warrant units.

> It was just the beginning. Customers in both countries would now have access to a seamless intelligent network.
>
> – BERT ROBERTS

Drexel had recently completed two large offerings of bond-warrant units for MGM/United Artists and Golden Nugget (now Mirage Resorts), and the market was extremely receptive to this type of security. So Milken strongly encouraged MCI to take advantage of the marketplace and double the size of the offering. McGowan and English agreed. MCI's $1 billion offering would be the largest nonutility corporate underwriting in American history, almost $100 million more than an offering AT&T had made two years earlier.

One reason why Milken wanted MCI to double the size of its offering—and why MCI acquiesced—was that it provided clout. When competing against a force as formidable as AT&T, MCI needed to show its customers, suppliers, and the marketplace that it had some muscle of its own. The extra $500 million would be proof to the world that MCI was in for the long haul. And of course, having excess capital never hurt a company.

Drexel's sales and research activities were truly impressive when focused on any offering, but the size of this deal inspired them to new heights. Not only must good investment bankers reach a wide circle of institutional investors, but they must provide the kind of detailed research and analysis that these sophisticated investors demand. Selling MCI's story meant helping investment managers truly understand the company and their plans for the future.

The day the registration was filed with the SEC, English was tied up with jury duty. At the end of the day, he rushed back to the office and called Milken. When English asked about the size of the book, referring to the amount of firm orders the salespeople had received from institutional investors, English got a surprise. "Over

The Offering Terms

Each unit consisted of one 9 1/2 percent, ten-year bond ($1,000 principal amount) and, adjusting for subsequent stock splits, seventy-two warrants that would allow the holder to buy shares of common stock. The bond portion was priced at $795.57 to yield 13 1/4 percent at maturity. The warrants were priced at $2.84 and were exercisable at $13.75 a share, a 31 percent premium over MCI's closing stock price on July 28 of $10.50 a share.

This offering was special not only because it provided value to the investor, but because it provided unique value to MCI. The bonds matured in ten years, and the warrants were good for five years. Since the stock price did not eclipse $13.75 during that period, however, the warrants expired unexercised—but not valueless. Every one hundred warrants were converted into one share of common stock. That allowed the expiration to be both nontaxable and of some value to investors.

The warrants increased the value of the offering for MCI because they allowed it to issue bonds at a lower interest rate, thus saving millions of dollars. And that was not all: By offering a combination of bonds and warrants, only $800 million was logged on the balance sheet as debt, even though $1 billion in cash was raised. The warrants, which were valued at $200 million, went on the books as permanent equity. Had this been an issue of convertible bonds, the entire $1 billion would have been logged as debt.

This is how a financier creates value for a company, by coming up with creative offerings that provide extraordinary benefits to both the investors and the issuer. Milken was an expert.

$2 billion" was Milken's answer. English was overwhelmed and elated. The efforts of the sales team and MCI's management had been rewarded with a huge oversubscription when demand turned out to be much greater than the offering. The $1 billion actually raised would be enough to ensure MCI's financial freedom for the next few years.

On July 29, 1983, Drexel Burnham and Shearson/American Express (the former Shearson Loeb Rhoades) priced the billion-dollar bond-warrant offering. The following day, the SEC declared the offering effective. It was a record-breaking event and an exhilarating time for all involved.

Milken made the deal happen and even appear seamless because the institutional securities buyers respected and admired his expertise and ability to produce quality offerings as well as his personal belief in, and commitment to, such offerings. Drexel and Milken did not walk away from a company or a transaction if the credit quality failed, and clients appreciated that commitment. If a company fell on hard times (as MCI did shortly after this offering), Drexel still made a market in the security, following the credit and arranging meetings between investors and the company. If something needed to be done, such as an exchange offer, Drexel did its best to bring it about.

The billion-dollar offering was creative, record-breaking—and exhausting. More than two thousand copies of the prospectus were mailed to potential institutional buyers and, remarkably, the buyers materialized in less than a week. (Drexel's years of marketing MCI to the investment community paid off.) But the pressure of pricing such an offering and confirming the orders is intense. With most new issues you cannot be sure that investors won't renege until you actually collect the money on the settlement date.

Because scores of institutional investors were participating, the book was nearly fifteen ledger pages long. Many of the buyers were money managers who bought for multiple accounts, resulting in thousands of tickets that needed to be processed. To make matters worse, Drexel still hadn't completely switched to electronic wire transfer. To say the pace was frenetic would be a considerable understatement. Although later transactions would be larger, none would be more memorable.

On August 8, 1983, MCI received a check for $988 million (net of underwriting fees); the issue was sold out. The offering,

> I can't think of anyone who got into trouble from borrowing. They usually get into financial trouble because of what they do with the money.
>
> MICHAEL MILKEN

On August 8, 1983, MCI received a check for $988 million.

MCI: Failure Is Not an Option

The Drexel Team

Drexel Burnham Lambert set standards for the high-yield market in the way it dealt with its clients. Much of what the company did is still having an impact in the 1990s even though Michael Milken is no longer actively engaged in the industry. His legacy lives on in the many former Drexel employees who now work at large investment banking firms and at the small, specialized boutique firms. They are carrying on the Milken tradition.

A lot has happened to me personally in the intervening years, but I am still extremely proud to have been a part of the historic offering and of the team effort involved in bringing it about. By team effort I don't just mean Milken's team, but all ten thousand Drexel employees, from the operations department to the corporate-finance and trading departments. Each of us had a role in MCI's financings. We believed at the time that we were doing something important—and history shows that we were.

Whether a company fails or succeeds, raising capital and giving entrepreneurs the opportunity to achieve is what our free market system is all about.

---

which was expected to cover capital needs for three to five years, raised almost as much as all of the company's previous public offerings combined.

This cash infusion came at a critical point in MCI's history. The antitrust decision had set in motion a chain of events whereby telephone users would soon be allowed to choose their long-distance carriers. If MCI could not build the technological and management infrastructure needed to take advantage of that onetime opportunity, it would languish. "If you had to point to one thing in the history of the 1980s that gave us the catapulting opportunity to go after that market," said Bert Roberts, "it would be that billion-dollar financing."

Did MCI Have Too Much Money?

At the time of MCI's $1 billion bond-warrant offering, some analysts claimed that MCI was adding too much debt to its balance sheet—accumulating too much money. An article in *CFO* magazine entitled "Why MCI's Conway Hoards Cash" painted an unflattering portrait of the company's financing efforts.

> More than two hundred institutional buyers (pension funds, savings and loan institutions, money managers, and credit companies) participated in the MCI offering including Fidelity Management, General Electric Credit Corporation, Cigna, Kemper, Northwestern Mutual, Columbia Savings & Loan, and Stein Roe.

Conway was unmoved by the criticism because he believed that a key to any company's success was getting financing ahead of its requirements. "The future is unknown and unknowable. But what you do know is the present. You know you can get money now. What you don't know is whether tomorrow, or a year from tomorrow, you'll be able to get money," Conway explained. "Get it while you can and get extra."

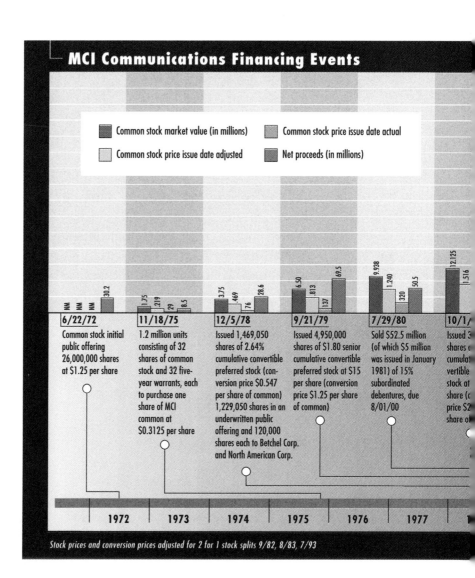

MCI: Failure Is Not an Option

Conway's philosophy is one that Milken has always stressed to his clients, and one that he urged MCI to adopt. Milken knows as well as anyone that the financial climate can change very quickly. Markets can go up and markets can go down. Investors will alter their objectives based on events in those markets.

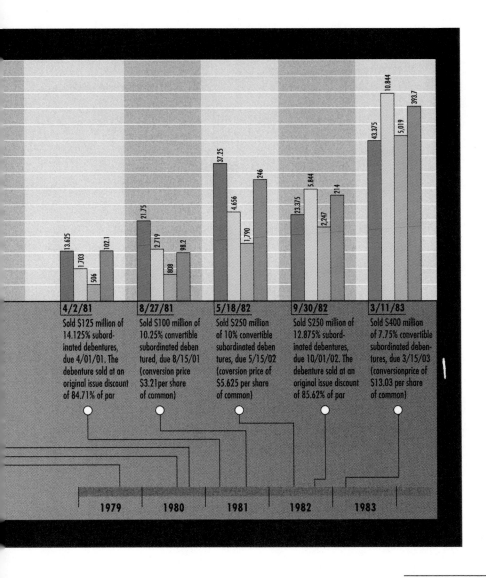

4/2/81 Sold $125 million of 14.125% subordinated debentures, due 4/01/01. The debenture sold at an original issue discount of 84.71% of par

8/27/81 Sold $100 million of 10.25% convertible subordinated debentured, due 8/15/01 (conversion price $3.21 per share of common)

5/18/82 Sold $250 million of 10% convertible subordinated debentures, due 5/15/02 (coversion price of $5.625 per share of common)

9/30/82 Sold $250 million of 12.875% subordinated debentures, due 10/01/02. The debenture sold at an original issue discount of 85.62% of par

3/11/83 Sold $400 million of 7.75% convertible subordinated debentures, due 3/15/03 (conversionprice of $13.03 per share of common)

1980–1984 ☏ The Growth Years

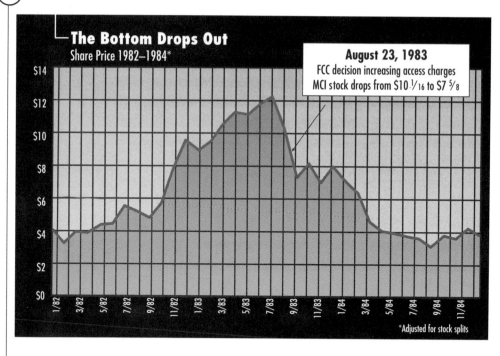

The Bottom Drops Out

MCI's stock took a beating after the FCC decision that was expected to double access charges paid by carriers like MCI. With the ruling, MCI's window of financing opportunity had shut tight and would not open again until the company came back into favor in 1986. MCI stock would not return to 1985 levels until 1988.

Unfortunately, no one could have predicted the magnitude of the change that was just around the corner. Three weeks after the billion-dollar offering, the FCC released an interim decision on the access charges that specialized common carriers paid to local phone companies; the decision was expected to double the charge to between $400 and $500 per line. MCI's stock price not only stumbled, it fell over and passed out. "Reaction was terrible," English said bluntly. MCI's stock dropped 26 percent in a day.

With this new FCC ruling MCI's window of opportunity for easy financing had shut tight. But fortunately for MCI's new investors, the structure of the bond-warrant unit protected their investment from the fallout. In fact, the bond-warrant unit outperformed other MCI securities. The unit as a whole lost only 6 percent of its value, dropping to $940 from $1,000.

Almost all the warrants, however, did expire unused because the stock price never reached the $13.75-a-share exercise price, and the price of the warrants themselves fell dramatically. The silver lining for MCI was that it was able to preserve equity. Moreover, it had the financial flexibility to weather the storm.

Although MCI management points to the bond-warrant offering as the key element that allowed it to survive and flourish, its ability to obtain financing is only half the equation. "I can't think of anyone who got into trouble from borrowing," Milken said. "They usually get into financial trouble because of what they do with the money."

MCI wisely put the money into the corporate infrastructure rather than using it to solve short-term problems. "MCI was like a developing country in the 1970s and 1980s," Conway said. "By that I mean, you look at some of the developing countries in the world today and they're trying to borrow money from the World Bank and other places. And the question is, what are they going to do with the money?" If countries use borrowed money only to buy food that will be consumed immediately, they haven't invested in anything that will lead to self-reliance and real improvement in standards of living. They would be better served to invest in seed grain and improved farming techniques. Similarly, companies that do not use financing to promote long-term objectives are not using it wisely; they are not adding value.

Chapter 5

Expansion and New Competition 1984–1990

> People who are part of the organization are also people who have an innate desire to change, who thrive on it.
>
> —BEVERLY POPEK

AT&T—Drawn and Quartered

As the world welcomed in the new year of 1984, Ma Bell officially waved good-bye to most of its corporate offspring. MCI's dream was now reality. But this watershed event in American corporate history had not been a foregone conclusion after MCI's antitrust victory in 1980. It was a stepped-up investigation by the Justice Department's Antitrust Division combined with pending antitrust suits filed by telecommunications competitors Litton Industries and ITT Corp. that finally made AT&T officials decide that spinning off the local operating companies would be less costly than trying to hold the conglomerate together. Unease among stockholders and the continuing costs of litigation in time, energy, money, and opportunities lost finally tipped the balance toward divestiture.

Having recognized the futility of continuing the fight, AT&T officials announced an agreement with the Department of Justice on January 8, 1982, that required the company to divest itself of twenty-two local Bell operating companies worth an estimated $80 billion—almost two-thirds of AT&T's assets. As part of the agreement, AT&T would keep Bell Labs, Western Electric, and its Long Lines Division, which handled long-distance communications traffic, and was also allowed to enter the fields of data processing, computer networking, and computer terminal sales, all of which were previously off-limits.

The twenty-two operating companies were to be reorganized into seven regional "Baby Bells" to provide local telephone service. These independently operated companies would then offer the same network access on the same terms to all competing long-distance carriers. It was this divestiture plan that was implemented on January 1, 1984.

As part of equal access, all residential and business telephone customers in the United States received a ballot giving them the opportunity to select a personal long-distance carrier.

Although MCI's wish had finally come true, the company would also learn that the fruits of AT&T's divestiture would not always be sweet.

As part of "equal access," all residential and business telephone customers in the United States received a ballot giving them the opportunity to select a personal long-distance carrier. This election process gave MCI a shot at grabbing a share of the $45 billion domestic market. In addition all carriers got circuits of equal quality, and MCI customers no longer needed to dial twenty-two numbers to reach someone in a distant city. (At long last McGowan could retire his joke about how you could identify an MCI customer by the callus on his dialing finger.) Furthermore, callers would no longer need a touch-tone telephone to access the MCI network, which meant that MCI could now compete for customers who owned rotary-dial telephones, almost half the country at the time.

But the scales were still tipped in AT&T's favor. The inertia factor alone gave AT&T a huge advantage: For many people it

was simpler to elect to stay with the name they had always known than to make a conscious effort to switch to another provider. Moreover, those who made no selection at all were assigned long-distance carriers by default. For the first eighteen months of equal access, the Baby Bells allocated almost all default traffic to AT&T, ignoring the other competing carriers.

After an FCC ruling in June 1985, however, the Bells had to allocate customers who hadn't yet made a choice based on the percentage breakdown among those who had. Each carrier, therefore, would receive the same share of the default group as it had won among customers who had actively selected a long-distance provider. This arrangement, of course, heavily favored AT&T because of its continuing overwhelming market share—90 percent in 1984. AT&T apparently was still a little more equal than the other carriers.

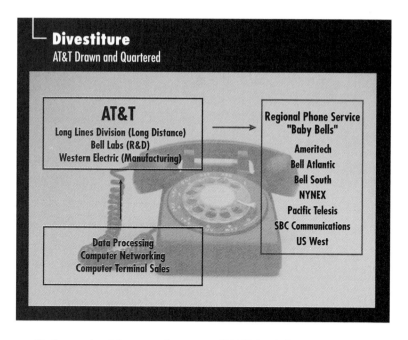

Perhaps the biggest advantage AT&T had, however, was the same one it had always had—money. As the long-distance carriers scrambled to gain market share, the hoopla generated by their competing campaigns reached a fevered pitch. With its deep pockets, AT&T was able to spend about $400 million on long-distance marketing in 1985, dwarfing the $50 million that MCI and Sprint each laid out.

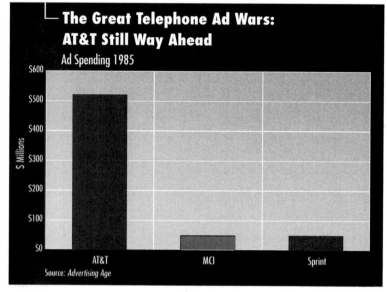

After the smoke had cleared from the long-distance voting in 1985, the combined top three companies had spent more than $500 million in a media blitz. AT&T still dwarfed its rivals, accounting for more than $400 million of the total.

As if the sheer size of its advertising budget wasn't daunting enough, Ma Bell also had the benefit of reams of data amassed during its decades of monopoly service, data that allowed it to fine-tune its pitch for specific customers. As the president of Sprint, Charles Skibo, complained to a *Time* magazine reporter in 1986, "AT&T had the data to sharpshoot and pick off select targets. We were shooting in the dark with a scattergun."

MCI got another nasty surprise when, in the wake of AT&T's divestiture, the rates it had to pay the Baby Bells (by now known as regional Bell operating companies or RBOCs) for local telephone interconnections rose sharply. The new rates were the first in a series of increases that eventually propelled the connection charges to astronomical levels.

MCI Decentralizes to Remain Competitive

AT&T's first stab at competing in the new environment came in May 1984 when it lowered its long-distance rates across the board by 6 percent. In June MCI countered by lowering its rates by 6 percent, allowing it to claim that its rates were still 40 percent below those charged by AT&T.

But just responding to AT&T was no longer enough. To keep its competitive edge, MCI decided to decentralize. In January

> AT&T had the data to sharp shoot and pick off select targets. We were shooting in the dark with a scatter gun.
>
> – CHARLES SKIBO

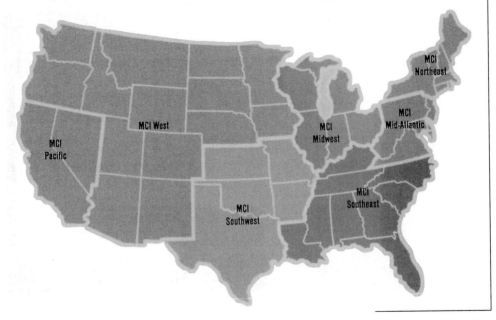

1985 at the urging of Bert Roberts, who since 1983 had been running the long-distance operating subsidiary, the company broke its telecommunications division into seven regional units covering roughly the same territory as the Baby Bells.

MCI assumed that with the end of the centrally controlled Bell System, each regional company—NYNEX, Bell Atlantic, U.S. West, Pacific Telesis, SBC Communications, Ameritech, and Bell South—would develop its own method of doing business based on local conditions. To deal with the regional Bells most effectively, MCI thought it needed to treat them individually, hence the separate divisions. In 1985 the regional Bells were valued at approximately $72 billion, with individual valuations ranging from $5.4 billion for Pacific Telesis to $16.2 billion for Bell South.

Each MCI unit was to be run as a separate company responsible for its own marketing, operations, and service, and the presidents of each had full profit-and-loss accountability. "You need that responsibility so that you push yourself, let the juices flow, feel the gas pains in your stomach," McGowan explained in his colorful style. The new system also let each unit experiment and explore, enabling the companies to take risks without affecting the rest of MCI's operations.

Even after the divestiture, MCI was still dwarfed by the size of the Baby Bells.

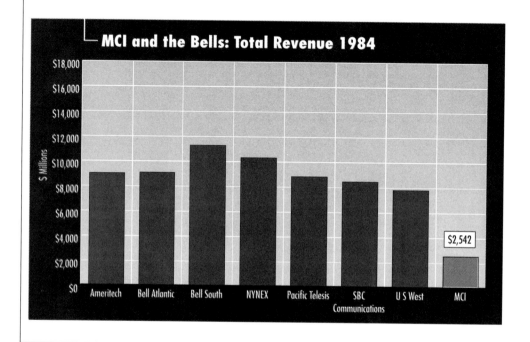

As COO Orville Wright had observed when the reorganization was announced in late 1984, MCI was seven times the size it had been in 1981. So by dividing the company into seven parts, Wright pointed out, "each new division is starting out where the entire company was three years ago."

This split, though it caused some managerial problems by draining headquarters of key personnel, proved MCI's commitment to keeping the company's spirit entrepreneurial. "Many premier companies take missteps. They get big. They get bureaucratic," Bert Roberts told *Forbes*. With the decentralization plan he designed for the company, Roberts was making sure that this didn't happen to MCI.

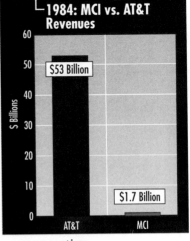

Among the most important leaders to leave headquarters as part of the decentralization was Jerry Taylor, who was dispatched to Denver. And while it was a loss to the Washington office, Taylor's presence gave the Western division the shot of adrenaline it needed.

"Jerry used to call our offices an auto-parts warehouse," Tim Price, president and COO of MCI, remembered of the Denver operation. "We were operating out of a big barn of a room with boxes of promotional materials everywhere and everyone shouting to one another. It was chaos!" But much like a newsroom or a trading desk, this kind of forced togetherness in a fast-paced, sometimes chaotic atmosphere creates real camaraderie.

In fact, at the end of the day people didn't want to go home. It became a tradition of sorts for Taylor and his crew to have dinner, drinks, and conversation together at Charms, the nearby watering hole. "We felt that we were creating something brand new, and Jerry's leadership was so powerful," Price recalled. "It was a magical atmosphere."

Helping to make things magical was Taylor himself. He is a manager with a generous spirit who encourages people to do their best, not so he can gain accolades, but so they can feel a sense of accomplishment and contribution. His management style, or non-management style as he sometimes characterizes it, is one that provides great latitude, but also expects great accomplishments.

Taylor believes that when people see that within their own sphere of influence their contributions matter, they become both happy and productive. "People don't get burned out from working hard. They get burned out from spinning their wheels, not getting traction. In some of the most stressful jobs, where people really put in long hours, if they can see the benefit, see things happen because of their labor, it invigorates and reinforces them."

Over the course of his career, Taylor has mentored many individuals who are the lifeblood of MCI today. They include Price; Angela Dunlap, executive vice president of corporate communications; and John Donoghue, senior vice president of consumer marketing, to name just a few. Small wonder then that when Taylor eventually left Denver in May 1987 to take on what was the company's biggest operation, the Mid-Atlantic division, his crew practically went into mourning. Price remarked that it was impossible for anyone to fill his shoes.

MCI's ability and willingness to make such pervasive structural alterations might seem hard for an outsider to understand. But management has never instituted change just for the sake of it, despite the company's reputation for aggressive change. "Anytime we have made a change, it's been for a good business reason," explained Beverly Popek, MCI's human resources director. And most MCI employees understand this philosophy; it's precisely why they're attracted to the company. "People who are part of the organization are also people who have an innate desire to change, who thrive on it," Popek adds.

> People don't get burned out from working hard. They get burned out from spinning their wheels, not getting traction. In some of the most stressful jobs, where people really put in long hours, if they can see the benefit, see things happen because of their labor, it invigorates and reinforces them.
>
> – JERRY TAYLOR

A Disappointing Conclusion to the Antitrust Wars

MCI's $1.8 billion award from 1980, which had been hung up in the U.S. Court of Appeals, was finally settled in May 1985. Ironically, what was once a major victory turned into a stunning disappointment for MCI. After five weeks of testimony, a jury reduced the award from $600 million to $37.8 million, which after being tripled for damages came to $113.4 million—roughly one-sixteenth of the $1.8 billion originally anticipated.

MCI's executives were devastated. They believed that the original award had been just and that this appeals court jury had been sympathetic to their cause. (Some jurors would later explain, however, that MCI's testimony was too confusing and technical.)

Never one to concede defeat, McGowan put the best possible spin on the situation by telling reporters that "the award of $113 million plus legal expenses is a substantial amount of money." As Roberts later recalled, however, privately the feeling was "total and absolute deflation." The plane ride back from Chicago to MCI's Washington headquarters was a quiet one.

McGowan refused to wallow in self-pity, though, and turned his attention instead to regrouping his troops. An appeal of the award, he knew, was still an option. "But," he told them, "you have to accommodate to what things are and move on."

In November, MCI, AT&T, and the regional Bell operating companies reached an agreement on both of MCI's antitrust suits. Though publicly MCI had been raising the ante in damage claims, privately it was looking to settle both suits out of court. With AT&T undergoing divestiture and MCI focused on wooing customers during the equal-access conversion, it made more business sense to spend money on competition than on legal fees. Apparently, AT&T thought so too. The final payment to MCI was estimated at just under $250 million.

MCI and AT&T would engage in other legal skirmishes in the years ahead, but there would be no more bruising, all-out legal battles. After eleven years of protracted conflict, the antitrust wars were finally over.

MCI and IBM Join Forces on Satellite Business

Amid the flurry of activity surrounding the corporate decentralization and the startling reduction of the antitrust award, MCI had been preparing to make another move that would stun industry analysts. In June 1985 the company announced an agreement to acquire the assets and operations of Satellite Business Systems (SBS) from IBM. In exchange IBM would receive nearly 47 million shares of MCI common stock, giving it a 16 percent ownership interest with the option of later expanding its holdings to 30 percent.

The investment community viewed this deal as a major coup for MCI, whose market performance had not lived up to the thunderous success of its billion-dollar offering. The stock had plummeted from just above $13 before the 1983 offering to less than $4—a 70 percent decline. The market was clearly cheered by the announcement as the stock rallied to above $5 again. (MCI gained 67 percent for the year versus 24 percent for the S&P 500.)

Although Drexel had been confident of MCI's prospects throughout the decline of the stock price, the IBM alliance provided a welcome sense of vindication. For an underwriter, which builds its reputation based on companies and the value of their securities, watching a high-profile client like MCI exhibit lackluster performance was disappointing. But this prestigious link served to justify the capital-raising efforts of the previous two years.

The IBM agreement was nothing short of wondrous. At the time IBM was the largest company in the United States, with a market value of $93 billion, more than twice that of the next largest company, Exxon. MCI's market value was puny in comparison at less than $4 billion. IBM was the bluest of the blue chips.

The transaction, valued at $465 million, was an important coup for MCI. McGowan believed that MCI would be more successful with SBS than IBM had been. Most importantly, however, he knew that

the relationship with IBM would give MCI an enormous boost in prestige.

The deal was also attractive to IBM because it was looking to get out of the telecommunications aspect of its business as it believed that telecommunications was becoming a commodity. SBS had been an unprofitable proposition for IBM, losing $120 million three years in a row. This unprofitability did not dampen MCI's interest, however, because of other advantages gained in the acquisition. Besides bolstering MCI International's relationships with foreign governments, most of management felt the association would be a key to helping win new customers, particularly in big corporations where MCI was at its weakest.

Although Conway and English both understood this reasoning, they felt that giving away 16 percent of the company (worth more than $400 million based on MCI's 1985 market capitalization) was a lot to pay for an "association." They also feared, despite IBM's assurances to the contrary, that IBM had plans to take over MCI. "Once you open the door to IBM," English told McGowan, "IBM may come marching through."

McGowan was unmoved. He described the IBM deal as "strong enough to break through the consciousness of most every businessman in the United States." How could it not? IBM's market value was twenty times that of MCI. (Ironically, by 1996

In 1985 MCI reconfigured its headquarters staff, creating a four-person Corporate Office that would have responsibility for strategic and operation management. The Corporate Office members were, from left to right, Bert Roberts, Bill McGowan, Orville Wright, and Richard Liebhaber, former senior vice president of engineering and development at SBS.

MCI  Profile

Douglas L. Maine

Doug Maine joined MCI in 1978, and jokes that he's held every job in the entire billing department since that time. He has served in several executive positions including controller, senior vice president of finance, and president of a fourteen-state operating division.

Today, as executive vice president and chief financial officer, he is responsible for a wide range of operations including tax functions, accounting, planning, administration, and investor relations.

Since Maine became CFO in 1992, MCI's market value has nearly doubled. He has also distinguished himself by directing a balance sheet restructuring that has provided the company with nearly $100 million in pretax interest savings.

Maine also served on the Board of Directors of Concert, the joint venture between MCI and BT.

A native of Lancaster, Pennsylvania, Maine holds a bachelor's degree from Temple University and a master's of business administration degree from Hofstra University.

IBM was less than four times the size of MCI.) Besides, acquiring the satellite business was MCI's secondary goal. What it really wanted—what it needed—was IBM's cloak of respectability and strength.

"There was a credibility question on the part of the largest accounts [as to] whether they could trust their very large and complex networks to MCI," Roberts explained. "It gave us a reference," observed Doug Maine, currently MCI's executive vice president and chief financial officer. "We were able to say if IBM is utilizing MCI's service, then it ought to be good enough for you."

As it turned out, however, SBS was not much more than a retooled and renamed version of CML Satellite Corporation, the company MCI had helped start in 1971 as a potential addition to its microwave transmission network before selling it in 1975. And like CML, the idea's potential never fully materialized.

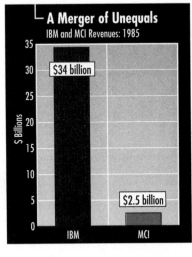

Although MCI had grown into a $2.5 billion company by 1982, it was still dwarfed by giant IBM at the time of their joint venture.

For numerous reasons SBS proved to be a difficult acquisition to digest. Financial struggles played a large role. MCI experienced losses in 1986 due in part to integrating SBS that forced the company to focus on other issues. In addition, IBM was experiencing problems of its own as the mainframe computer market stagnated and its personal computer sales failed to take off. IBM's stock price dropped from $147 a share on the day the MCI agreement was announced to $120 by the end of the year, losing 18 percent (about $17 billion in market capitalization) while the S&P 500 gained 65 percent. (Big Blue's share price continued to struggle in subsequent years, hitting a low in 1993 of $41. In July of 1996 it was trading at around $95.)

But the deal was strategic nonetheless because it served its primary purpose of putting MCI's credibility question to rest with large ac-

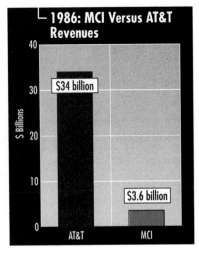

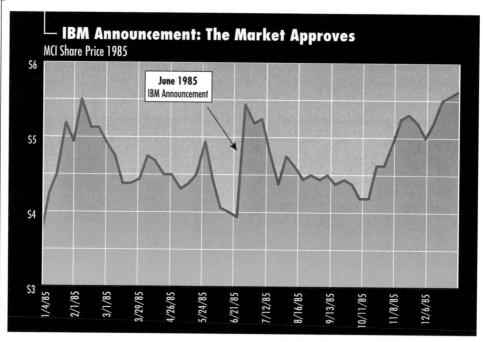

counts. "The relationship with IBM, or at least the perceived relationship," Roberts related, "was critical in the largest accounts believing they could do business with MCI, that we would be a credible vendor to provide them what they needed to compete in their industries."

In addition the IBM affiliation opened the door to the capital markets once again. One month after the SBS acquisition became final and less than three years after the billion-dollar underwriting, MCI returned to the capital markets and sold $575 million of subordinated debentures. (At this point, MCI was so comfortable with Milken's ability that McGowan hammered out the offering's details with him in a most informal way—at a lunch break in the middle of Drexel's High Yield and Convertible Bond Conference. The conversation took all of five minutes.)

The offering helped to improve MCI's financial picture in 1987, but the joint venture remained stalled. Finally, in 1988 an increase in corporate profits and an accumulation of cash reserves allowed MCI to buy back its stock and independence from IBM for $677 million while retaining the perception of a strong business connection.

Although MCI and IBM parted ways, the two maintained a friendly working relationship, and though the relationship is not quite as extensive as it was back in the mid-1980s, IBM is still an MCI customer.

MCI Sells Its Cellular Business

As part of its 1981 purchase of Western Union International, MCI had acquired a paging and cellular telephone business called Airsignal International. Although it was only a small part of WUI's business, Airsignal was the third-largest radio common carrier and paging company in the country. Its paging operation, including mobile telephone service, had 36,000 customers in thirty-nine states. Industry analysts correctly saw enormous potential for growth and profit in the paging industry.

MCI believed that Airsignal would fit nicely into its growth plans. McGowan, having read that executives were away from their telephones 90 percent of the working day, announced that "the pager market for business executives is one of the major untapped segments of the communications industry."

Jerry Taylor, who had been called back to Washington a couple of years before to oversee the entry into the residential long-distance market, was chosen to head the Airsignal subsidiary. He wasted no time in rolling out an advertising campaign in Dallas that touted pagers as "the greatest tool for business since the telephone." With competitors like Paging Network, Inc. and TDS's American Paging subsidiary hard at work to garner a share of the market, Taylor nevertheless promised that MCI planned to be "the company that leads this industry into the twenty-first century." By 1983 Airsignal had 76,000 pagers in service, bringing in $18 million in revenues annually.

But MCI's plans for Airsignal were shattered in 1984 when the FCC rewrote the rules for the booming paging industry, announcing that it would hold a lottery to award three national franchise licenses. As luck would have it, MCI came in fourth.

Although disappointed at being shut out of the paging market, MCI's management thought the future of mobile telecom-

MCI Communications and McCaw Cellular

The paths of MCI and McCaw Cellular have crossed many times throughout their life spans. Although their first direct involvement was MCI's 1986 sale of Airsignal to McCaw, the similarities between the two companies had fostered a cordial informal relationship. Ironically, the two companies were often mistaken for each other—McCaw Cellular Inc. had the same initials, they were both considered to be upstart companies in the telecommunications business, and they both were entrepreneurial in spirit and quick in making things happen.

Craig McCaw's admiration of MCI surely springs from its parallels to his own story. Like the founders of MCI, McCaw recognized the potential of an idea—in his case the cellular telephone—before anyone else did. He began with a tiny cable-TV company in Washington state that he inherited from his father, which he ran from his dorm room at Stanford University. When he recognized the potential of the cellular business, he began to buy and trade licenses, and eventually sold the cable business altogether in 1987. He raised a large amount of capital (including $1.25 billion raised through Michael Milken) and went on a buying spree. People called him crazy, but he saw what most people did not: that every license he bought could easily be sold to the Baby Bells at a substantial profit. Their belated recognition of cellular's potential allowed McCaw to become one of the most important names in the telecommunications industry.

His foresight proved to be flawless: McCaw grew into a hugely successful company, and its success was complete when it was purchased by AT&T for $11.5 billion in 1993.

Since that sale, McCaw has headed up Nextel Communications, with an eye toward developing its national digitial-wireless network. He is also working on a revolutionary new telecommunications concept—an "information skyway" on a system of 840 satellites that will surround the earth and free information from the confines of cable. His critics call the idea crazy, but McCaw once again looks forward to proving them wrong.

munications really lay in cellular phones, not paging, so the company refocused its energies on applying for cellular licenses. Once again, however, the FCC turned to a lottery to determine who would get the licenses (although this lottery was for the large markets only, not the entire nation), and MCI lost out again. These losses forced it to rethink its strategy.

The company concluded in 1985 that its holdings, including fifty individual paging markets and cellular franchises in seven

By 1983 Airsignal had 76,000 pagers in service, bringing in $18 million in revenues annually.

cities, did not constitute the critical mass necessary for real growth and profitability in either business. It decided to sell Airsignal to a subsidiary of McCaw Cellular Inc. for $116 million.

In hindsight some people might perceive this decision as a mistake. But the company's core business was long-distance telephone service. If it had used its resources—both human and financial—to break into cellular, it could well have jeopardized the service that buttressed the entire company. Orville Wright believed that the cellular market would drain capital that could be used to modernize its network without providing any returns for at least a decade.

This view was affirmed by Craig McCaw when he later commented on MCI's 1986 sale of Airsignal to his company. "Occasionally when you're flying the airplane," said McCaw, "you do have to throw some baggage off even if the baggage is very valuable." He also added that because MCI understood the importance of focusing on its core business, "it ended up the most successful competitive phone company in the world."

McCaw's evaluation has a familiar ring to MCI's management. This pick-your-battles theory is one to which Taylor wholeheartedly subscribes. "Jerry's whole philosophy is to focus on the big things, do them well, and let the little things go," explained

> We had that single-mindedness of focus in our cellular business that they had in the long distance business.
>
> – CRAIG MCCAW

1984–1990 ☎ **Expansion and New Competition**

Angela Dunlap. "At the time that we sold Airsignal, equal access represented a $45 billion opportunity. We needed every dollar, every good person working on that."

MCI's unique ability to keep its eye on the ball is one of the major reasons that the company has been able to stay in the game. "There are always opportunities, sometimes ones that are sexy or that appeal to certain groups of constituents," observed Dunlap. But management has consistently been able to concentrate on the biggest opportunities for growth.

Advertising Age named Dunlap to its Marketing 100 list, and the Financial Women's Association voted her 1995 Woman of the Year.

A Changing of the Guard

The year 1986 brought far more changes than just the truncation of MCI's cellular business. Just after celebrating his fifty-ninth birthday while vacationing in Virginia Beach in mid-December, Bill McGowan suffered a severe heart attack. (Coincidentally, three days later, John deButts, the former chairman of AT&T and McGowan's chief adversary, died.)

The tragedy sent shock waves through the company. McGowan's management team worshipped him. "We were like little children, and we totally trusted 'Dad'," commented Harris. "Everyone felt they were McGowan's special person, and we all were. It was true."

McGowan's illness came at a time when MCI was already in turmoil. Profits had been severely constricted by the jump in costs associated with equal access, the acquisition of SBS, and increased competition from AT&T and Sprint. In early December the company announced it would lay off 15 percent of its workforce and take a $585 million write-off, more than $450 million of that related to its outdated satellite and analog networks. After forty profitable quarters, MCI posted an operating loss of $448 million, the largest in its history.

These were dark days for MCI. Apart from dealing with McGowan's illness, having to lay off people was heartbreaking. The company had always prided itself on treating its workforce well. "He [McGowan] enjoyed the people that worked at MCI, no matter who they were and what level they were," said his wife, Sue Ling Gin. "They were all part of a team and there was a lot of camaraderie and loyalty."

Speculation abounded in the media, the industry, and the investment community as to the severity and duration of McGowan's illness and its potential effect on the company. Shortly after the announcement about McGowan, a *Washington Times* article pointed out the obvious: It was going to be a "blue Christmas" for MCI. That comment was relatively minor, however, when compared with the drubbing MCI received in a *Financial World* article in which MCI was named as one of the ten most poorly managed companies in America.

Despite the timing and public perception, the uncertainty surrounding MCI's business had little to do with McGowan's heart, and his illness left the organization far from rudderless. McGowan had already begun grooming Bert Roberts. By then Roberts was a seasoned fourteen-year veteran of MCI, and it was expected that he would take over as CEO and chairman. Furthermore, the actual management of the company had been the responsibility of Roberts since Orville Wright's retirement fifteen months earlier. (Roberts had become president and chief operating officer at that time.) In fact, it was Roberts who came up with the plan to decentralize the company in 1985 and Roberts who saw the need to restructure again along product lines in the fall of 1986. Roberts was competent, knowledgeable, tough when he had to be,

M C I Profile

Angela Dunlap

As executive vice president of corporate communications, Angela Dunlap is responsible for crafting and articulating MCI's corporate and grand strategy to key constituencies including investors, the media, legislators, and employees. Dunlap also oversees the definition and alignment of MCI's brand strategy across various market segments, and manages MCI's community outreach efforts.

In her fourteen years with MCI, Dunlap has served in a variety of positions including president of marketing for MCI Telecommunications Corp. and president of MCI Consumer Markets. Under her leadership MCI introduced Friends & Family Connections and Concert, the joint venture with BT.

Dunlap was named one of the "Marketing 100" by *Advertising Age* and was selected as the 1995 Woman of the Year by the Financial Women's Association.

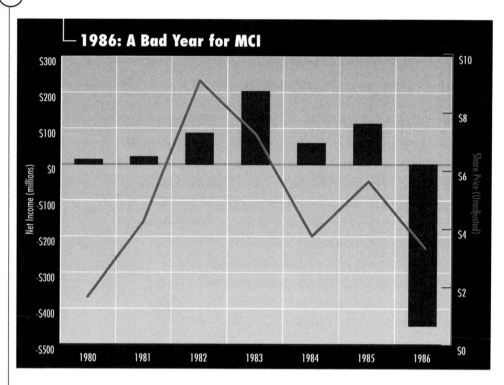

and, most importantly for MCI in these bleak times, would not accept defeat.

Company insiders knew that McGowan was more effective as a visionary than as a manager and that he had actually played a lesser role in daily operations since as far back as 1975 when Wright was named chief operating officer. Wright now was coaxed back to help run the company following McGowan's heart attack. McGowan had had the foresight to delegate areas of his authority. He did not want one person to be critical to the success or failure of the company.

McGowan may have symbolized a gutsy and successful MCI, but the financial setbacks plaguing the company were endemic to all long-distance companies in 1986 and were in no way related to his physical collapse. Massive layoffs, equipment and restructuring write-downs, and spending cuts were representative of the industry's competitive environment. During the 1970s and 1980s the industry experienced many changes caused by such factors as intensifying competition, the AT&T divestiture, and shifting technologies. MCI, AT&T, and other telecommunications companies were sailing into largely uncharted waters. This time MCI could not count on McGowan to provide navigation.

McGowan believed that, like many other heart attack patients, he could be back to work in a few months. Unfortunately, his condition did not improve. His wife recalled that he had had some tough days, but for the most part McGowan stayed positive. When the rest of the family expressed reservations about a heart transplant, McGowan, ever the lover of high technology, was all for it. In the spring of 1987, he received a new heart. Though he did eventually return to work part-time, McGowan would never again be able to devote the same level of energy to MCI.

Fiber Optic Network Goes National

MCI's coast-to-coast fiber optic network began service in January 1987, four years after the company purchased the more than 150,000 miles of fiber optic cable needed to build it. Laid primarily along railroad rights-of-way, the delicate cable had been expensive and slow to install, making the financing deals the company completed during this period absolutely critical. Although some analysts had speculated that MCI had gone overboard with its consecutive $1 billion and $575 million offerings,

Workers sew fiber optic cable into the ground along a railroad throughway.

MCI found that it needed every penny for the new network. Capital expenditures during the three-year period ending in 1986 were more than $3 billion. Without those two key offerings, the company would have been much more restricted during this stage of its expansion.

The system quickly proved its worth, carrying a large volume of traffic at high speeds and with extraordinarily high quality. The new cable greatly increased MCI's carrying capacity and enabled it to keep pace in the fast-changing industry. Because of its 1986 losses, however, MCI decided to postpone plans to build further fiber optic routes, opting instead to lease routes from Williams Telecommunications Company through 1987. By then, the financial picture was again bright enough for MCI to get back to building its own network.

In July 1988 the company announced construction of another fiber optic route between Houston and Los Angeles. At its completion in 1989 the 1,630-mile link gave MCI two complete transcontinental fiber optic pathways and provided the company with more capacity and greater flexibility. With its innovative application of fiber optics, MCI established new standards for the telecommunications industry, finally becoming known as a company that could compete in both price and quality.

The Price of Success

At the same time that fiber optics was giving MCI more capacity to carry calls, the company's profit per call had been going down. FCC-ordered price cuts for AT&T services had forced MCI to lower its rates in response. And equal access, which MCI had struggled so long and so hard to obtain, was now giving the company growing pains.

Before AT&T's divestiture, MCI's local interconnection charges were 70 percent below those paid by AT&T, based on the fact that MCI received inferior interconnection facilities. As equal access was phased in, however, MCI's access-charge advantage over AT&T got slimmer and slimmer.

So as competition drove down prices across the board—from 1984 to 1988 MCI lowered its rates 36 percent, an average of 9 percent per year—access fees were eating up an increasing portion of company revenues. In 1984, the year of the AT&T di-

By the mid-1980s MCI had invested billions of dollars to recreate its network, replacing an out-of-date microwave analog radio system with state-of-the-art digital and fiber optic technology. The map below charts the MCI network in 1987.

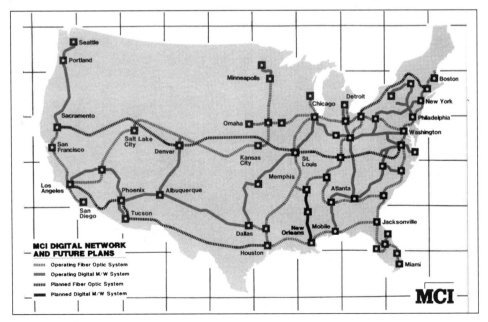

1984–1990 ☎ Expansion and New Competition

vestiture, MCI paid access charges to the local operating companies of $480 million, about 25 percent of its $2 billion in revenues. By 1988 MCI's access charges had rocketed to $2.4 billion, or 47 percent of its $5 billion in revenues.

But the fivefold leap in access charges was, in part, a reflection of the company's success: MCI's network carried 24 billion billable minutes of long-distance calls in 1988, an increase of 32 percent from the year before. By the end of 1988 traffic exceeded 100 million minutes a day.

MCI Blitzes Market With More New Services

Part of this increased traffic was due to the new toll-free 800 service that MCI introduced in 1987. The prepaid service allowed a business to set up a line that customers could call for free. In its first year of operation, the service became indispensable to mail-order firms, delivery companies, and businesses with customer call-in and assistance lines.

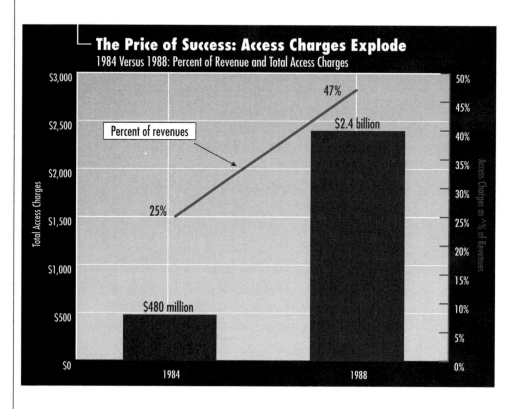

The Price of Success: Access Charges Explode
1984 Versus 1988: Percent of Revenue and Total Access Charges

Competitors such as Sprint and AT&T already offered 800 service, but MCI executives saw a niche. They also expected the multibillion-dollar market to grow at an annual rate of at least 15 percent. MCI's salespeople emphasized the company's simple, per-minute, distance-sensitive rates, which contrasted favorably with AT&T's cumbersome, more complicated pricing structure. One of the first clients to sign on was the U.S. Immigration and Naturalization Service, which awarded MCI a $1.3 million contract to set up toll-free lines to handle questions from the public. By the end of 1988 MCI had captured nearly 10 percent of the domestic 800-number market. With the added capacity of its fiber optic system, MCI was well positioned to seize a hefty share of this market.

MCI was aggressively expanding into other areas as well. The volume of its international calls surged from fifty thousand minutes per month in 1984 (in Britain only) to twenty-three million minutes per month in 1987, with service available to 80 percent of the world's telephones. MCI's corporate sales also took off with long-distance contracts from United Airlines, the Home Shopping Network, and even the Pentagon.

These successes helped propel 1987 earnings to $88 million or 31 cents per share (in contrast to a 1986 loss of 79 cents per share following the $585 million write-off). Moreover, MCI reached a milestone in 1987 by collecting $1 billion in revenue in the last quarter. At year's end, the company found itself with $725 million in cash, a strong financial position far removed from the cash-starved years of the 1970s and early 1980s.

MCI continued its expansion in 1988 by introducing operator services. Its customers previously had access to customer-service lines but not operator assistance to place individual long-distance calls. Throughout the rest of the year MCI phased in additional new services including assistance with collect calls, third-party billing, and international calls. The new operators not

> From 1984 to 1988 the long-distance market grew 22 percent from $51 billion to $63 billion. MCI's market share more than doubled, from 4.5 percent to 10 percent.

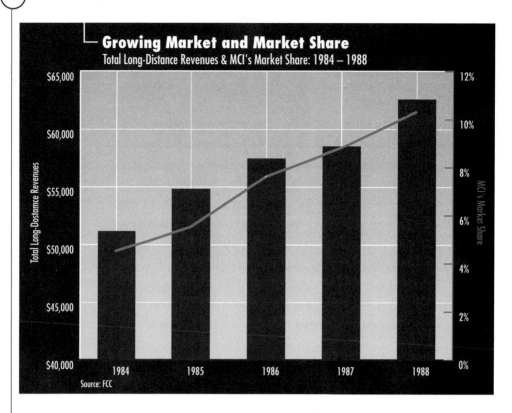

Growing Market and Market Share
Total Long-Distance Revenues & MCI's Market Share: 1984 – 1988
Source: FCC

only improved service, they kept customers from leaking back to AT&T for special calls, thus ensuring that MCI would earn every dollar possible from its subscriber base. By the end of 1988 MCI operators were handling over two million calls per month. Furthermore, the long-distance market had grown from an estimated $45 billion in 1984 when equal access was implemented to more than $50 billion four years later.

McGowan Gets Back Into the Groove With MCI Fax

MCI's competitors were confident that McGowan, with all of his health problems, was firmly entrenched in retirement. But less than three months after his heart transplant, McGowan strolled into his office unannounced. He was back—at least part-time. He would return full-time around Labor Day and assume the job of co-CEO with Wright.

He was welcomed back to Drexel's High Yield and Convertible Bond Conference in 1988 and, as anyone who knew him might have expected, he was ready with a quip. "I was not here last year, the first time since 1980," he said. "I had a hell of an excuse—but if you don't mind the pun—I did have a change of heart and am back this year."

McGowan most wanted to exert his influence on new services. That year, one innovation the company launched was a 900 service featuring two-way-conversation capability. AT&T's 900-callers, for instance, could only listen to prerecorded messages. The best new product, however, was of McGowan's own creation—a dedicated network for fax messages.

The idea was spawned by an ad McGowan had seen in the *New York Times*, placed by a New York City electronics store that was offering fax equipment. McGowan showed the ad to Tim Price, then vice president of general business marketing. "Those guys don't waste a nickel on anything that isn't making money," McGowan told Price, referring to the fact that the store was devoting increasing amounts of its advertising space to fax machines. McGowan wondered why MCI didn't have a product utilizing this technology. He sought out Price because he believed the fax market was hot, and he wanted someone who could move quickly. In fact, McGowan told Bert Roberts, "I want someone who will go around everybody in the organization to get this product out," Price recalled.

It was October, and Price courageously promised McGowan a product by the second week in November. "We began writing teaser ads in October for a product that we were inventing as we went along," said Price. Others in MCI management weren't quite as gung-ho as McGowan and were looking to take things more slowly. But McGowan did not want to wait this one out. His direct order to Price was simply, "Launch this thing."

The company introduced the new fax service through the MCI International subsidiary where it enhanced the existing MCI Mail service, an electronic-mail service begun in 1983. With the new product, users of both MCI Mail and telex services now could send faxes from their personal computers to any other fax machine without additional hardware or software. New support services included detailed monthly reports listing all faxes sent and received, the option to arrange for toll-free 800 numbers, and the ability to charge transmissions to credit cards.

McGowan offered a glimpse of the future by assigning the fax service a major role in MCI's plan to become a "one-stop source for global communications." MCI estimated fax services to be a $3 billion market and expected it to grow at a rapid pace. (Indeed by the mid-1990s MCI estimated that 50 percent of all calls made during the business day were fax calls.)

The Value of Human Capital—Getting Things Done

One of MCI's enduring strengths has been its human capital. More than its technology, more than its ability to get financing, MCI has derived its strength from its people. "We recognize how crucial our employees are to our success," related Roberts, "and we are firmly committed to developing, nurturing, and rewarding the talented employees who will achieve our goals."

MCI recognizes employee contributions in many ways. One of its recognition programs is the Spirit of MCI Award, which is given to those who have been able to make extraordinary things happen such as getting a big contract signed, coming up with profitable new ideas like 1-800-COLLECT, or achieving a remarkable feat of customer service. "What is valued most highly in this company—and I think this really sets it apart—are ideas," observed Bob Schmetterer, a partner at the New York-based firm that handles MCI's advertising. "The people most frequently rewarded are those who have had great ideas and have made those ideas happen."

One recipient of the Spirit of MCI Award was MCI operator Will Klausen, who had helped a little girl whose baby-sitter had collapsed. When the little girl, Kaitlin, called and said that her baby-sitter was asleep and wouldn't wake up, he recognized that something was wrong, traced the call, and dialed 911. He then kept the child on the phone, asking her about her favorite cartoon characters and generally entertaining her to put her at ease. When the police arrived, he made her comfortable enough to open the door.

Tim Price, who is now president and chief operating officer, is a prime example of MCI's commitment to rewarding people for extraordinary performance. When Price started working on the new fax service, he was by his own admission an underling about four layers removed from reporting directly to McGowan. Yet McGowan sought him out because he perceived Price as someone who could get things done. But he is only one of many who have been promoted from within the ranks. Roberts and Taylor are themselves veterans from the company's early years.

McGowan gave these people and others a shot because neither pedigree nor diplomas held any sway with him. What he valued—and what MCI management still values today—was creativity, hard work, and, most of all, enthusiasm for the job. At MCI enthusiasm is engendered by giving people autonomy and, as a result, a sense of worth. McGowan liked to joke that MCI had moved decision making so far down in the ranks that if a loading-dock worker went through three levels of management, he'd reach the chairman.

> People who are part of the organization are also people who have an innate desire to change, who thrive on it.
>
> —BEVERLY POPEK

1984–1990 ☎ Expansion and New Competition

"We delegate fairly important decisions very low in the company," commented Angela Dunlap, who is proof that MCI is also unequivocally gender-blind. In delegating decisions, however, MCI also requires employees to accept the responsibility of decision making. "We have this culture that rewards progress, not just activity," she explained. "It's not just sitting around in meetings talking, but making things happen." In fact, Taylor is so confident in his staff's ability to make decisions that he never even sees the company's advertising spots until they appear on television. "There's nothing I could offer," he remarked, "other than approval." Although management evaluates the strategy and the advertising message, how that message is delivered is entirely up to the advertising staff.

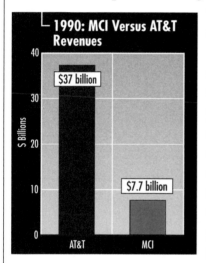

In some ways this hands-off approach is a natural evolution from the early days when McGowan and Taylor were decidedly hands-on, particularly with advertising copy. Stories of this pair designing ads and writing copy themselves are part of company lore. Both were instrumental in the decidedly unsubtle, but effective, early ads spoofing AT&T. Recognizing the value of such experience, Taylor wants to give younger employees the same opportunities, and he is confident they can do the job.

Confidence, however, is not the only factor determining Taylor's insistence on delegating power. He believes that the company needs younger people writing the copy and designing the campaigns because most of MCI's advertising is aimed at that population. "We need people with rings in their noses, ponytails, and red spiked hair who are giving direction to the copywriters," Price said. "We don't want a bunch of stuffed shirts if we hope to communicate with the market that is the future."

One of MCI's target markets is the segment of college-age young people through the upper end of middle-age baby boomers, which represents a large, mobile population that does not think twice about picking up the telephone to keep in touch with friends and family around the world. "We clearly want to be

MCI Profile

Timothy F. Price

Tim Price, president and chief operating officer of MCI Communications Corporation, is responsible for the company's North American operations, including operating alliances such as Avantel in Mexico and Stentor in Canada. He serves on the Board of Directors of Concert Communications Company and the Board of MCI Systemhouse.

Previously Price served as president and chief operating officer of MCI Telecommunications, the company's core communications business, where he led MCI's expansion into the highly competitive $100 billion local market. Price continues to drive MCI's growth in the U.S. local market where the company plans to offer local service in thirty-one major markets by mid-1997.

Since joining MCI in 1984, Price has been instrumental in shaping MCI's reputation as a company that seizes new opportunities in being first to market with innovative products and services. Under his leadership, MCI introduced the industry's first integrated communications brands—MCI One for consumers and networkMCI One for businesses. Most recently, Price unveiled MCI's latest integrated technology initiative, VAULT architecture, which combines traditional telephone networks with the Internet, once again demonstrating the company's pioneering use of advanced technology.

Price's marketing expertise has earned him wide respect both inside and outside the industry. He was recently named Outstanding Corporate Strategist by Frohlinger's Marketing Report, Marketer of the Year by *media-week magazine,* and Marketing Executive of the Year by the National Accounts Management Association.

perceived as hip, fun, current, and a part of pop culture," Price explained. "And we want to be known as the company that brings people closer to the people they love the most."

MCI thinks of this market as the ATM generation, because according to a Gallup poll conducted in 1990 people who use ATM cards have a high propensity to be MCI customers. "If you think about it, there are two lines at the bank every Friday afternoon," added Price. "There's a long line of people waiting for the teller. Then there's a proportionately much shorter line of people waiting at the ATM machine. The people in the teller line don't want to use the machine because they're afraid of technology. The younger people in the ATM line tend to view technology as a friend."

Over the past ten years as more and more people have become accustomed to ATM machines, the bank lines have changed considerably. MCI reasons that this trend will only accelerate, and as ATM use grows, so too will MCI. If MCI is to keep pace with this population segment's needs and desires, the younger generation's contributions are crucial.

Chapter 6

The Dawn of a New Age 1990—1992

AT&T still thinks that MCI has done to long-distance telecommunications what Attila the Hun did for urban renewal.

— BILL MCGOWAN

MCI Focuses on Advertising and Marketing Strategies

Despite its phenomenal growth during the 1980s, MCI did not leap lightly into the last decade of the millennium. Having invested heavily in equipment and technology in the early part of the decade, even when capital was limited and often at the expense of short-term profits, MCI found itself sitting on top of an underutilized state-of-the-art fiber optic network. Management had to figure out how to most effectively use what it had struggled so hard to build.

With network construction and legal battles for the most part behind it, MCI had to refocus on advertising and marketing. It needed to concentrate on telling people about the company's services. "No matter how well you do something," McGowan would say, "you need to tell people about it." MCI also always believed you needed to be as audacious as possible in the telling. But, as the company had learned some years before, it wasn't always acceptable to just go over the top. In 1983, for instance, CBS had refused to run a spot developed by Ally & Gargano for MCI Mail that poked fun at the United States Postal Service. The message needed to be both brazen and cleverly packaged.

MCI learned to communicate directly to both private and business consumers through advertising.

MCI's first ads to beef up its residential business in 1989 trumpeted its prime time long-distance plan, which began discounting calls at 7:00 p.m. local time, three hours earlier than AT&T's competing Reach Out America plan. At MCI's press conference to announce the product, Price declared that it was no longer necessary to "reach out and wake someone." Price was delighted to see his words quoted on the front page of the next day's *Wall Street Journal*. The ads themselves packed the same kind of wallop. Reminiscent of MCI's first residential television spots from 1980, they displayed a split screen with two people sitting side by side, calling the same long-distance location. At the bottom of the screen, two meters ticked off the cost of each call, with the dollars and cents zipping by on AT&T's side while mounting more gradually on MCI's side.

AT&T Fights Back With Advertising

1990 was the year that AT&T introduced their "Put It in Writing" campaign, which challenged MCI to "put its money where its mouth was." The campaign cut MCI's residential sales in half in ninety days and slowed its revenue growth to 0 percent by the fourth quarter of 1990. MCI's stock took a nosedive.

This ad, however, was only the first salvo in what would become a bitter and expensive advertising war. Shortly after the MCI ad appeared, Sprint, the nation's third-largest telecommunications carrier, joined the battle, referring to Reach Out America as "Rip Off America." AT&T struck back in 1990 with its "Put It in Writing" campaign, driving home the message by telling MCI customers (in a six-million-calls-per-month telemarketing effort) to have MCI "put its money where its mouth is." The long-distance war deteriorated into a no-holds-barred brawl, eventually spilling over (as usual) into the courtroom, where AT&T and MCI accused each other of deceptive advertising practices.

AT&T was adopting MCI's dramatic style of advertising. It even mentioned MCI by name. "That was a real admission that the market had moved," observed advertising executive Tom Messner. AT&T was abandoning its consumer-friendly ads to get

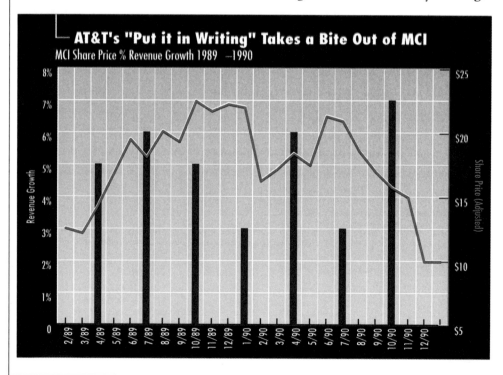

MCI: Failure Is Not an Option

down and dirty. The "Put It in Writing" campaign turned Ma Bell into more of a Ma Barker, and in so doing announced to the world that MCI was no longer a pesky upstart but serious competition. So serious, in fact, that at the time of the "Put It in Writing" campaign, MCI was taking 100,000 customers away from AT&T every week through telemarketing.

The ad campaign hit hard. Although MCI had always put its savings in writing on its bills, it was more difficult to prove those savings to non-customers. By forcing MCI's hand—making it send proof in the mail—AT&T effectively slowed down the company's marketing efforts. "It gave a phrase for people to get the telemarketer off the phone," Price recalled. "It implied that telemarketers were nuisances that should be dealt with that way. It managed to imply that we were not telling the truth."

The "Put It in Writing" campaign was a real blow to the motivation of the MCI workforce, which had been upbeat, even euphoric since AT&T's divestiture. The successful marketing strategies that had brought it into the mainstream of the industry and allowed it to capture increasing market share had promoted the feeling that it was a real player and not just a fly-by-night novelty. That self-concept, however, was damaged when AT&T decided to use its muscle and play advertising hardball.

"The effect of advertising isn't just on the consumer marketplace," Schmetterer explained. "It has tremendous impact on morale inside the target company. When you're in a very sales-focused business, which this is, morale has a great deal to do with how successful you are." With AT&T attacking and morale dropping, MCI's success began to wane. Residential sales dropped by 50 percent in ninety days.

Despite AT&T's short-term victory, however, the negative advertising from both sides eventually backfired. Consumers were uneasy with the endless round of hostile and confusing advertising claims and soon became apathetic. When the smoke cleared, it was apparent that nothing much had changed except the public's perception of the industry. Nobody won great numbers of new customers, and the whole industry had lost face.

> AT&T still thinks that MCI has done to long-distance telecommunications what Attila the Hun did for urban renewal.
>
> – BILL MCGOWAN

1990–1992 ☏ The Dawn of a New Age

One of the things that did change, however, was the way MCI handled its sales calls. It decided to step up its quality assurance with third-party verifiers to combat the consumer apathy that was a residue of all the negative advertising. These verifiers called every telephone sale MCI made to make sure that the new customers understood the terms of their contract and agreed to any switching costs that might be involved. Although MCI instituted this control of its own accord, when the FCC eventually decided to regulate the industry's telemarketing in December 1992, it used MCI's system as a template.

Another benefit that came from AT&T joining the street fight was the way MCI was viewed in the marketplace. By specifically naming MCI in its ads, AT&T officially made it a two-horse race. AT&T thereby not only recognized MCI as a worthy competitor, but implied that MCI was its only competitor.

"In binary decisions," remarked Price, "the number two guy always wins disproportionately." If AT&T had simply said, "We're better than everybody else," MCI would have been lumped in with Sprint and other competitors. But by singling MCI out, AT&T was advertising the fact that it considered MCI to be the only other real choice for consumers.

Friends & Family Is Born

When Friends & Family was introduced in 1991, MCI was at a crossroads. AT&T's "Put It in Writing" campaign had knocked the wind out of the company. The campaign had taken a huge bite out of MCI's residential sales and had demoralized MCI employees. Furthermore, MCI had fallen out of favor on Wall Street. The stock slid from an all-time high of nearly $24 in October 1989 to less than $10 a share by the end of 1990.

"We needed something big," remembered Price. "We were going to be out of the residential business if we didn't find something magical."

To pull the consumer market out of its nose-dive, Bert Roberts turned once again to Jerry Taylor, then-president of the Mid-Atlantic division. Taylor was tapped to head up the company's entire consumer business. At the time of Taylor's appointment, each division had been running its own consumer segment. Now all consumer business would be centralized under one person, the first time since the company's overall decentralization. Taylor, in turn, recruited Tim Price, who by then was vice president in charge of general business, to help him in this new mission.

"Jerry came to me in the fall of 1990 and told me he had been asked to run consumer markets for the whole company," Price recalled. "As a matter of fact, we discussed the issue three times over lunch, and Jerry kept talking about how he'd need somebody to do sales and marketing, somebody that he could feel comfortable with."

Price was interested in the job and believed he had subtly indicated his interest to Taylor more than once, but apparently he had been a little too subtle. Finally, he said, "Look, Jerry, if you want me to recommend somebody else, I will, but I would be delighted to do this myself. I hope you'll consider me." According to Price, Taylor looked at him and replied, "Well, all I've been looking for was some expression of interest." With that, Price was on board to help create the discount calling program that has been hailed as one of the most brilliant marketing initiatives ever.

Taylor and Price lived and breathed consumer marketing for nearly three months. "We did everything as a team," Price reported. "We both had to learn about it together." And just as employees had congregated in Denver after work, the Friends & Family team often spent their evenings at the Ritz bar in Washington, having dinner, exchanging ideas, and sharing information. "It became almost a regular staff meeting down there," Price remembered. "If you got too busy to discuss something with someone during the day, you could go down to the Ritz and find them—no matter who it was or what position they had—and find them ready to answer your questions."

Taylor has an explanation for the closeness that forms among MCI employees. "We've never been able to do a lot of things at one time until recently—not by design, but by necessity," he observed. "So we would get people to rally around one mission. Whether it was Friends & Family, building the system, whatever, that one thing gave focus. And when you have people working at it, contributing to something that they can see and touch that is a success, it's wonderfully exhilarating for people."

The mission for Taylor and Price this time was to build a new service that would play to two of MCI's strengths—telemarketing and technical capability. The company already had the technical capability to bill calls placed between MCI customers at differing rates based on perceived relationships between those customers, and its telemarketing force, of course, was second to none. The pair agreed that it was important to offer significant savings, and the idea at first was to charge a onetime fee to offset the hit the company would take to revenues. But Taylor knew from his Execunet days that customers weren't fond of extra charges. Price's answer was a referral program: The lost revenue could be balanced by increasing the customer base. Thus was born the concept of Friends & Family.

The service combined MCI's advanced network technology with creative marketing. Using its sophisticated computer system, MCI could search through its entire database of customers, which had been accumulated one by one, and qualify a preselected calling circle to receive discounted rates. In effect Friends & Family customers had custom-designed, private networks.

Another plus for MCI was that its customers effectively became MCI employees! "It's a pricing methodology and a way to turn your customers into salespeople," Bert Roberts told *AdWeek* magazine in a September 1995 interview. By having customers provide a list of twelve people whom MCI could ask to join a calling circle, the plan was, as one industry analyst said, "a good, cheap way for someone else to do quick, cheap marketing for you." Management knew that if each current customer brought in just one new customer, MCI's customer base would grow by 100 percent.

Particularly ingenious was the fact that with Friends & Family, MCI created a product category in which AT&T couldn't compete, a perfect example of corporate jujitsu, where the

> We would get people to rally around one mission whether it was Friends & Family, building the system, whatever, that one thing gave focus.
>
> – JERRY TAYLOR

weight of the market leader is used against it. Because AT&T had 80 percent market share at the time, most people included in the calling circles would already be AT&T customers. If it offered a competing 20 percent discount, AT&T would be offering that discount on eight of every ten calls made. Because MCI was looking to expand its market share, it could make the numbers work.

Brilliant Ideas and Quick Management Decisions

The product was brilliant, and Roberts recognized it immediately. When Taylor and Price made their presentation to him in January, he listened attentively as they explained why they thought it would work during a presentation that took hours. The meeting was long, but productive—Roberts asked the important questions and Taylor and Price had all the answers. Finally, at the end of the meeting, Roberts asked, "When would you be ready to launch?" "March," they replied. "Get it done in February," Roberts instructed.

> Management knew that if each current customer brought in just one new customer, MCI's customer base would grow by 100 percent.

Just like that Roberts had made a decision that would change the course of the company forever. Price wonders how many other CEOs would or could make that call without some kind of committee review. "I remember walking out of there," he said, "thinking this is how business should operate." It was quick and cool. That's what makes MCI different.

Roberts took decisive action just as a good leader should, but that doesn't mean his actions were rash or hasty. He knew full well all that was at stake: If Friends & Family flopped, MCI would not only be out of the residential sector, but its business market could be threatened as well. As McGowan had pointed out at an MCI sales meeting several years earlier, the two services are firmly linked. "No matter how hard we service our national accounts, when a top executive goes home, he's just another consumer. We can either lose or find ourselves in jeopardy on an account because a client can't get a clear connection to his or her mother or can't get a $12 overcharge resolved."

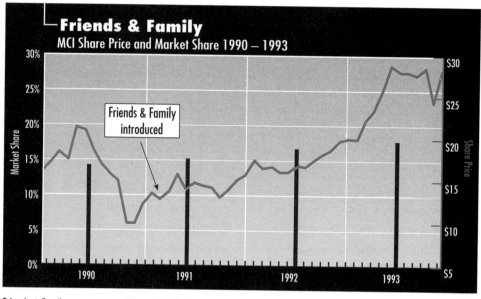

Friends & Family gave a jumpstart to growth in share price and market share.

Though Roberts gave Friends & Family the go-ahead, he also recognized that it had an Achilles heel—systems support. The ability to provide the correct discounts to the correct people in all the calling circles was system dependent. To allay Roberts's concerns and remedy the problem, Taylor and Price developed a management information system (MIS) department specifically for the consumer division.

Friends & Family was a go, and as MCI had done so many times before, it moved quickly forward without looking back. Taylor and Price had been stretching it when they told Roberts they could launch in March, and now they had instructions to make it February. Their solution to this dilemma was another brilliant stroke of marketing savvy—teaser ads.

By mid-February MCI had ads on the air featuring Zsa Zsa Gabor and George Burns naming people who would be on their lists, without ever saying exactly what those lists were all about. Zsa Zsa said her list would include her ex-husbands and the police officer she slapped (a highly publicized incident that had landed her in jail). George Burns said his list would include various female bridge partners in cities across the country and delivered the punchline, "Well, at my age I can still play bridge."

The ads were funny, memorable, and they piqued consumer interest—so much so that MCI actually had people calling to sign up before the program was fully explained! Within ten weeks of its actual launch on March 18, 1991, Friends & Family had added one million new customers. By January of 1993, it had ten million. Along the way the Friends & Family campaign, from the original home-movie-style spots that evoke nostalgia for the 1950s to the more recent ads in which Whoopi Goldberg makes a young friend part of her calling circle, has left an indelible mark on the nation's consciousness.

MCI knew it had done its marketing job well when Jay Leno mentioned the service in some of his "Tonight Show" monologues. In one bit in which Leno joked about the capture of famed terrorist Carlos the Jackal, he said, "You know how they caught him, really? Apparently, Yasir Arafat had Carlos' phone number on his MCI Friends & Family plan. Just dialed it."

With Friends & Family, MCI was not just bringing in new customers but working to change public perception. Part of the calling plan's strategy was to soften the company's sharp and aggressive image. In one memorable public relations event, Bert Roberts accepted an invitation to attend the wedding of a Maryland couple who wanted to thank MCI for letting them court inexpensively. Not only did Roberts show up at the wedding, he took along a camera crew, and the footage was turned into a Friends & Family commercial.

Although marketing to families is popular now, it was not commonplace in 1990. MCI had research done showing that people felt as if they had lost touch with their families. It indicated that people moved an average of five times before they were forty. What made Friends & Family such a powerful concept from an emotional perspective was that it combined the notion of staying in touch and MCI's reputation for saving customers money.

"All of a sudden we were the company that connected you to people you cared about," said Angela Dunlap.

When Friends & Family added its ten-millionth customer, MCI celebrated by offering one free month of service to new

> Within ten weeks of its actual launch on March 18, 1991, Friends & Family had added one million new customers. By January of 1993, it had ten million.

customers. Within two years of the program's launch, the company's share of the U.S. long-distance market had expanded from 14 percent in 1990 to 17 percent, all thanks to Friends & Family. And the company's revenues leapt from $8.4 billion in 1991 to $11.9 billion in 1993—a 41 percent increase.

Part of the reason Friends & Family was a hit is that it was the first program of its kind. The notion of moving forward quickly and being first to market is one of MCI's primary tenets, Dunlap remarked. "Even if it's not perfect, you're in the market first." This strategy helps solidify brand identity in the consumer's mind.

A Marketing Breakthrough

Friends & Family was also the first telephone calling plan to actually have a personality, an extraordinary feat considering that it isn't a product per se, but a service. The phrase "friends and family," for instance, replaced the previously common "family and friends" in most Americans' speech patterns. In fact, in a survey MCI conducted, it found more people knew about Friends & Family than knew that Al Gore was vice president of the United States or that Hawaii is a state (sad, but true).

According to Bob Schmetterer, Friends & Family was the first of several revolutionary breakthroughs in marketing. "For the first time, long-distance calling really moved from a price-oriented commodity to a brand—something people could grab on to. Before this product, there were company names—AT&T, MCI, Sprint—but never a brand name that meant something unto itself." Although other companies, including AT&T, have tried to come up with brand names for telecommunications products, no one has been as successful. "Over the last six years, MCI is the only company I know that has actually developed meaningful, consumer-based brands," Schmetterer argued. "If you ask anybody to tell you about Friends & Family, they may say bad things about it or they may say great things about it, but they know what it is."

AT&T's response to Friends and Family was characteristic of the company's ongoing advertising brawl with MCI. Ads for its heavily promoted plan called "True USA" implied that the so-called true savings with Friends & Family were not what MCI represented. Never one to turn the other cheek, MCI retaliated with a classic spot. It featured an AT&T executive hooked up to a polygraph machine, while an investigator asked whether AT&T had exaggerated the improved quality of its new True Voice system. The executive's troubled answer, of course, was yes.

MCI responds quickly to competitor claims not only to set the record straight, but also to save its customers from embarrassment or doubt. "Our customers are emotionally connected to us," Tim Price remarked. He believes they want the company to fight back. "Our customers have made an investment in MCI and they feel like they are being personally attacked."

The company also uses its competitors' false charges to help instill a sense of mission among its employees. "What helps us is that we have an enemy to focus on," Price added. "And the enemy, God bless them, is so vicious, so big, and so formidable that the meaner their advertising gets and the more mean-spirited their public comment gets, the more enthusiastic our people become."

Growth Around the Globe

Although the domestic advertising wars and the resounding success of Friends & Family were grabbing the major share of the spotlight (and press attention), MCI had not closed its eyes to opportunities elsewhere. With its acquisition of Western Union International back in 1982, MCI had gained 22 percent of the overseas telex market. But, as the decade drew to a close, that market was rapidly being squeezed by newer, higher-speed data transmission methods (primarily the fax machine) and by personal telephone calls. To further its plans to become a dominant international carrier, therefore, MCI acquired RCA Global Communications, Inc. (Globcom) for $160 million from General Electric in May 1988.

The benefits of this acquisition were substantial. Globcom had a solid customer base and plenty of contacts abroad, particu-

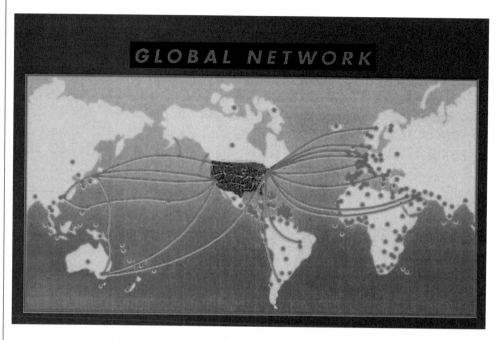

GLOBAL NETWORK

> Going global is not cheap…[but] failure to go global could be very expensive indeed.
>
> – BILL MCGOWAN

larly in Asia where MCI had only a small presence. The acquisition also enhanced MCI's position in its struggle with the European postal, telegraph, and telephone ministries (PTTs) to open new markets. Moreover, Globcom was part owner of several new transatlantic and transpacific fiber optic cables.

By acquiring Globcom's combination of customers, government contacts, and technology, MCI gained the additional tools it needed to become a major force in the international voice and data communications market. By the end of 1988 the company offered telex, leased-line, voice, and electronic-mail service in 150 countries. MCI could now market itself as the nation's only broad, full-service international carrier.

Just as its parent company had done domestically, MCI International revolutionized the overseas telephone industry by restructuring monopolistic foreign markets into competitive ones, lowering the cost of communications, and stimulating industry growth. As one PTT official put it, "Before, the international telephone business was a gentlemen's club; now, it's a business."

Continuing its push to expand internationally, MCI joined with Telefonos de Mexico early in 1990 to provide direct voice service between the United States and Mexico. Direct connections were vital if MCI was going to hold down its international rates. (A few years later, in 1994, the Mexican connection was strengthened when MCI formed a joint venture with Banamex-Accival to build a long-distance network in Mexico. Seeking to take advantage of the exploding market for communications up and down North America, MCI agreed to invest $450 million for a 45 percent stake in the new company, called Avantel.)

In addition MCI and its European partners started laying the groundwork, quite literally, for an expansive international telecommunications system. In 1990 they began building the world's largest undersea fiber optic cable, called the TAT-X link, between North America and Europe. Eighteen telecom companies from eleven countries shared the $400 million cost of the project. When TAT-X was completed in 1993, it could handle 150,000 calls simultaneously and provided five times the capacity of its predecessor, the TAT-9, in which both MCI and AT&T also held part ownership.

Meanwhile, plans for other transatlantic cables were on the drawing board. AT&T and Deutsche Telekom were looking to build the TAT-G-1 between Germany and North America, while Sprint and Cable and Wireless plc of London envisioned a successor to their jointly owned PTAT-1, which had been the first of the privately operated transatlantic cables.

MCI further increased its international presence by buying 25 percent of the outstanding shares of INFONET, a world leader in electronic data transmission. As the largest shareholder of INFONET, MCI's stake brought with it additional international revenues and strategic partnerships with ten major international telecommunications providers.

All this activity resulted in expanding international telephone traffic over the entire industry. Figures for 1991 showed total industry growth of 13 percent. In 1991 MCI ranked number four in this market. It was behind AT&T, British Telecom, and Cable and Wireless, but was experiencing the strongest growth by far—up 35 percent that year alone.

> By acquiring Globcom's combination of customers, government contacts, and technology, MCI gained the additional tools it needed to become a major force in the international voice and data communications market.

A Mighty Merger

Although 1990 was a landmark year for MCI both stateside and abroad, perhaps its biggest step toward continued growth was its $1.25 billion acquisition of Atlanta-based Telecom*USA.

The merger, the largest in telecommunications history up to that time, was part of Roberts's strategy of "getting big fast." Now well on his way to becoming MCI's CEO, Roberts was exerting his influence. Although he had always been satisfied to work in McGowan's shadow, insiders knew Roberts to be equally aggressive and feisty.

Financed through a $1 billion loan from a Bank of America-led consortium of twenty-two banks, the acquisition of the fourth-largest long-distance telephone company represented a significant change in course from MCI's earlier strategy of increasing market share by pursuing large accounts and by targeting residential users with aggressive advertising.

The benefits from the acquisition were many. Telecom*USA, a strong company with annual revenue of $750 million, brought with it a half-million new customers in thirty-one states, three thousand miles of fiber optic cable in the Southeast and Midwest (territories where MCI had little or no fiber optic cable), a different mix of customers (primarily small and midsize companies),

Telecom*USA

Atlanta-based Telecom*USA was formed in 1988 in a merger of two smaller regional carriers with O. Gene Gabbard at the helm as CEO. (He moved to MCI with the merger). It quickly grew into an industry leader in the fast-growing business of voice messaging and other services for executives on the road.

By the time of the merger with MCI, it served 500,000 customers and generated sales of more than $700 million. Telecom*USA had been aggressive in developing new products, enjoying particular success with a credit card that allowed holders to make long-distance calls from most pay phones, send and receive recorded messages, and set up conference calls. MCI was particularly interested in this area as the long-distance market became more competitive. Telecom*USA also owned a 3,000-mile fiber optic network radiating from its central locations in Georgia and Iowa.

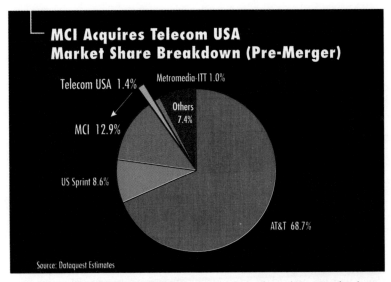

and one additional percentage point of market share in the long-distance market. The acquisition also prevented another company from buying Telecom*USA and becoming a formidable competitor. The number of major U.S. long-distance carriers would remain, therefore, at three for the time being until a series of mergers and acquisitions brought WorldCom, Inc., the fourth-largest long-distance company in the United States, into the game with about 5 percent of the market by 1996.

With the acquisition, MCI bought 1.4 percent of market share.

A Legend—and Leader—Dies

McGowan, who continued as chairman of the board after Roberts became CEO, retained his corporate warrior image until the end. As he presided over MCI's 1992 annual meeting in May, the usual recitation of annual meeting business was interspersed with McGowan's humor. When asked by a stockholder at one point when the company's stock price might return to its previous highs, he replied, "You have to wait until the markets open. It might be selling there now. You can't tell."

But even McGowan's unflagging spirit could not prevent his exhausted body from finally calling it quits. On June 8, 1992, while waiting to begin an exercise

The *Los Angeles Times* paid tribute to business innovator Bill McGowan upon his death.

class for transplant patients at Georgetown University Hospital, McGowan collapsed and died of heart failure.

Although his death was not totally unexpected, it stunned the business world, particularly those who knew the man personally and had benefited from his visionary approach to business. McGowan's dogged persistence and enthusiasm for life were admired by all and created a sincere devotion in the people whose lives he touched. Mike Milken paid a visit to McGowan while he was recuperating after his transplant. McGowan welcomed him even though he was too sick to see many others, and though he was so weak that he could barely get out of bed, McGowan still found excitement in little things. Always delighted by new technology, he was playing with a hand-held copier machine when Milken walked into the room. Milken and McGowan had become good friends over the years, and weathered their share of storms together. McGowan actually found time to keep a scrapbook with newspaper clippings about Milken, which Roberts presented to him after McGowan's death.

Throughout the years McGowan had channeled his zest for life and his irrepressible determination into transforming MCI from a small, struggling company with under $1 million in sales into one of the most successful corporations in the U.S.

But perhaps most importantly, Bill McGowan was the champion and father of today's free market in long-distance communication. From 1984 to 1996, the price of an average long-distance call plummeted by almost 70 percent. He radically transformed an entire industry and, in so doing, enriched society. MCI gave consumers and businesses a choice in long-distance service. It provided wealth for investors, suppliers, and bankers, and it created jobs. Today, MCI employs more than 55,000 people.

AT&T's chairman Robert Allen graciously acknowledged McGowan's impact on the industry after his passing: "Probably more than any other single person, he helped to reshape the long-distance business from the monopoly that it had been for so long to the highly competitive industry that we know today."

In a warm and humorous tribute to McGowan's competitive nature, his friends at the company's New York ad agency designed a memorial poster that hangs in many offices around MCI's Washington headquarters. It reads: "If there's no competition in heaven today, you can bet there will be plenty tomorrow." People didn't just admire McGowan for his business prowess, though, they connected with him on a personal level. If you asked around, you would find dozens of people at MCI who considered McGowan a good friend.

Portrait of the American Dream

"Most people told me it couldn't be done," Bill McGowan once said. "Others even said it shouldn't be done—that AT&T was a natural monopoly and it was against the public interest to have competition in telephone service."

McGowan, of course, scoffed at such notions. He battled in court to give MCI equal access to the nation's telephones. Opponents warned that more competition would mean higher costs, because there would be fewer customers to support the enormous investment and operating expenses of each competing telephone network, and lower quality, because less income per company would translate into poorer service to customers.

In the end McGowan proved the naysayers wrong. After the 1982 consent degree forced AT&T's divestiture and MCI claimed full access to local telephone systems, American consumers for the first time were able to choose their long-distance provider. Prices dropped and quality improved as a result of the new competition, exactly the opposite of what critics had predicted. Competition taught all phone companies—even AT&T—how to do more with less and do it better.

Till the end of his days Bill McGowan remained a champion of competition and a catalyst for change, radiating the spirit of a true pioneer.

As devastating as McGowan's death was to his family, friends, and colleagues, Wall Street barely blinked an eye. The Monday after McGowan's death, MCI's stock closed at $32.875, up 62.5 cents a share. The fact that the market did not react was a sign that the transition of power to Roberts and his team would be a smooth one.

In a press conference following McGowan's death, Roberts, who knew he had the confidence of his troops, still chose to

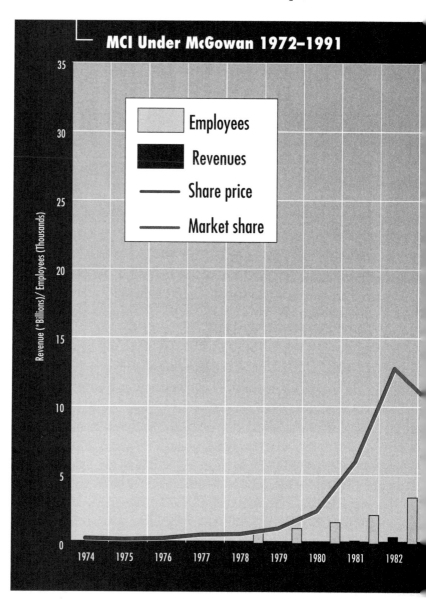

allay any unspoken fears. "We're known for our tenacity, entrepreneurial spirit, and can-do attitude," he said. "I want to make sure that these will continue to be our trademarks."

On June 23 the board voted to name Bert Roberts chairman of MCI. "Everybody felt safe," recalled Price, "because he was the guy who had been running the company." Despite the loss of its leader, MCI moved on.

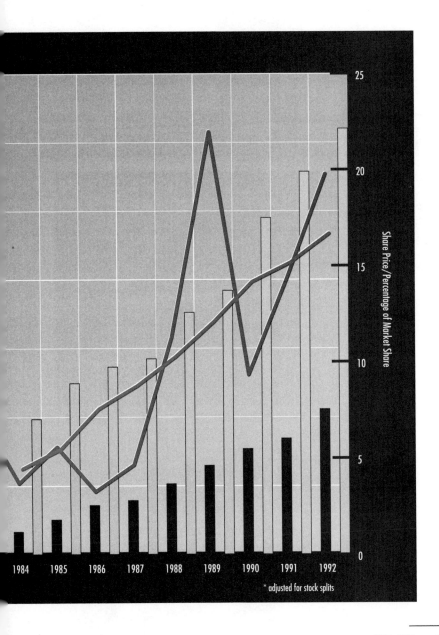

Chapter 7

New Directions 1992—1996

By its steadfast refusal to accept change, to accommodate change, or to admit change, monopoly guaranteed change.

—BILL MCGOWAN

Digital Conversion Complete

For MCI, moving on meant bringing to completion the long-term projects begun by Bill McGowan and associates, forming partnerships with others in the industry, and harnessing the many new markets created by advances in technology.

After writing off its outdated analog microwave network, MCI completed the conversion to digital transmission in 1991. With its entire network operating digitally, MCI had become what the head of the company's systems engineering group called "a big computer with long wires." (AT&T also had been working to upgrade its network, and could report by the end of 1990 that 99 percent of its calls were digitally switched although it had set no deadline for completion of the changeover.)

Digital technology provides cleaner and faster transmission of both voice and data. Rather than replicating sound waves mechanically, as in the analog system, digital transmitters send a stream of pulsed codes that are regenerated at the receiving end. A major advantage is that digital transmissions do not weaken over distance. Although the change to digital transmission was expensive for MCI, most analysts agreed that it was a smart long-term move.

MCI headquarters at 1801 Pennsylvania Avenue NW in Washington, D.C.

MCI called its digital system the "intelligent network" and exploited its potential by introducing new products such as MCI Preferred, a small-business discount plan, and Global Communications Service, which made business services available to companies all over the world. MCI's resourceful use of technology again kept it a step ahead of the competition.

New Advertising for a New Era

About this time the company's advertising began to take a new tack, one that reflected the technological changes sweeping the industry. Bob Schmetterer credits this new direction directly to Jerry Taylor, who he believed had the "clarity of vision" to realize that the future would not be about long distance, but about something much, much bigger.

With the company looking to develop its network infrastructure and build it up to handle the digital future, Schmetterer and partner Tom Messner worked together with Taylor, Price, and Dunlap to find a way to articulate Taylor's vision. The result was two ad campaigns that positioned MCI as a leader in next-wave technology and progenitor of the future. The first campaign was a remarkable series of ads featuring child actress Anna Paquin—who later won an Academy Award for her supporting role in the 1992 movie *The Piano*—in which the eleven-year-old talks in visually stunning settings about the future wonders of the Information Age. The ads created a stir and associated MCI with the cutting edge of information technology.

MCI's advertisers now set about providing examples of these advances in a second ad campaign called "Gramercy Press." In this series of commercials viewers are brought into the lives of sympathetic characters at a fictional publishing company who struggle with, and are enriched by, new technology. "From an advertising standpoint, it was a revolution," Schmetterer explains, "because it became a series, a kind of soap opera, in which the characters became known to the public." The ads featured the characters talking about their company and how MCI was working with them to put e-mail in place, provide Internet access, and ultimately take the company into the twenty-first century. Their story of adapting to these powerful technological tools and services was one almost everyone could relate to personally.

Indicative of the revolutionary nature of the campaign was the publicity it generated. Messner and Schmetterer were asked to appear on the ABC television program "Good Morning America" to talk about the campaign and introduce some of the characters to the television viewing audience. It became front-page business news, and *USA Today* ran a series that tracked its progress. By talking about the future of telecommunications in terms of people and their everyday lives, rather than in terms of

MCI's Five Marketing Breakthroughs

Bob Schmetterer, who came on board as part of the MCI advertising team in 1990, has taken a close look at how MCI has revolutionized marketing in the telecommunications industry. His analysis has enabled him to pinpoint five specific events that he considers to be breakthroughs for the industry.

1. The Friends & Family program transformed long-distance service from a price-oriented commodity to a brand name that represents something tangible to consumers.
2. The "Anna Campaign" was one of two ad campaigns that positioned MCI as a leader in next-wave technology and progenitor of the future. In a series of commercials child star Anna Paquin raises questions as to how technological change will affect the telecommunications industry. This segment was the first of two ad campaigns that Schmetterer groups together as "Defining the Future."
3. The second segment of "Defining the Future," called "Gramercy Press," was intended to answer some of the questions the Anna Campaign brought up. It generated an amazing amount of publicity on television and in newspapers by taking viewers into the lives of sympathetic characters at a fictional publishing company who struggle with and react to the changes technology is bringing to their traditional industry. The ads ran in serial format complete with sub-stories, slice-of-life vignettes, and cliff hangers—a truly new direction in the field of advertising.
4. Approaching the business market as a mass market rather than a niche market enabled MCI to communicate its message more powerfully to this highly profitable segment. Up to this time advertising for business products generally meant print ads in business journals such as *Fortune* or *Business Week*. Soon everyone who wanted to appeal to business customers approached them as consumers through mass media.
5. The 1-800-COLLECT program, which took an old idea like collect calling and created a whole new image for it solely through advertising, enabled customers to call collect from any telephone anywhere regardless of the long-distance carrier.

All of these events unfolded during the period from 1990 through 1996, which Schmetterer considers "the most creative, explosive, unbelievable" time in the company's advertising history.

digits and processing speeds, Messner and Schmetterer took the yawn factor out of technology and made it interesting to a broad-based audience.

Schmetterer labeled the Anna and Gramercy campaigns "Defining the Future," and so successful was this whole approach that it was carried over into the consumer market. A series of commercials called the "MCI Connections" revolved around a family scattered across the United States whose members communicate news about the new baby's name and other family matters via e-mail. The impact of this media blitz was that the common perception of what communication was and what it could do for us had changed forever.

A Major Victory in the Air

In 1992 MCI garnered two more important successes. While MCI was setting out to define the future of telecommunications, the Federal Aviation Administration was reassessing the nation's air-traffic-control system. Unhappy with the old system's level of reliability, the FAA wanted tower-to-pilot communications and radar information as well as access to critical computer data. In March 1992 it awarded MCI a $1 billion contract to upgrade the system. The contract was a major victory for MCI not only because of the lucrative compensation but because of the prestige of being entrusted with such a critical system.

MCI developed the Leased Interfacility National Air Space Communications System, or LINCS. It consisted of digital fiber optic and microwave transmission components designed around an intelligent switching platform. LINCS, which had a backup for every communication path, established new industry standards for reliability and technical service.

An Old Idea With New Life—1-800 Goes Portable

Later that year came yet another opportunity for MCI when the FCC ruled in favor of 800-number portability. A business could now use its 800 number with more than one long-distance provider at a time, in effect carrying the number from one provider to another. The ruling stripped AT&T of an advantage that had arisen because some businesses were reluctant to switch long-distance companies for fear of confusing customers with new 800 numbers.

To take advantage of this opportunity, MCI developed a farcical, but nonetheless effective, public relations campaign. It fired up its employees and grabbed new customers with the hook that AT&T had been "the warden of the 800-number prison" for the past twenty-five years, but finally customers were free to scramble over the prison walls and choose their preferred carrier. The idea was picked up and used over and over again in print, giving MCI a lot of publicity as well as business.

MCI celebrated Billion Dollar Day March 3, 1992, having saved America $1,000,000,000 compared to AT&T's prices.

In another classic instance of marshaling its troops, MCI enlisted the aid of its employees by asking them to record any 800 number they saw—on billboards, refrigerator magnets, product packaging—wherever. From this effort MCI collected an astonishing 300,000 different 800 numbers. It then used these leads to contact potential customers and invite them to switch their 800 business to MCI.

This campaign, of course, led to yet another advertising volley. AT&T responded with ads that called into question MCI's 800 reliability. MCI returned serve with spots featuring bar charts that depicted MCI successes relative to AT&T.

The ads may not have won any creativity awards, but for MCI they served a distinct purpose: The company was hoping to goad AT&T into retaliating with similar advertising. By getting AT&T

to compare itself to MCI, management figured the public would think that there were only two options for 800 service—even though Sprint as well as a number of smaller competitors were jumping into 800 service.

AT&T took the bait and soon began running ads that showed MCI's relative market share. As with its "Put It in Writing" spots, AT&T was singling out MCI as the competition. Although the ads were designed to show how minuscule, and thus ineffective, MCI was in relation to AT&T, they had the opposite effect. "All it did was reinforce our growth," observed Price. "People would say, 'I had no idea MCI was so big.'"

Proof Positive

The year 1992 was also a big one for MCI in the business market. MCI's marketing team quickly recognized that the most important obstacle in the market was sheer confusion. If everyone claimed to be cheaper, obviously someone was not telling the truth. It was up to the Atlanta-based business-markets division, which Price had taken over in 1992, to help customers sort the whole thing out.

In a typically lightning-quick rollout—ninety days from conception to market—Price and his crew came up with a product that offered businesses three proofs that MCI was their best

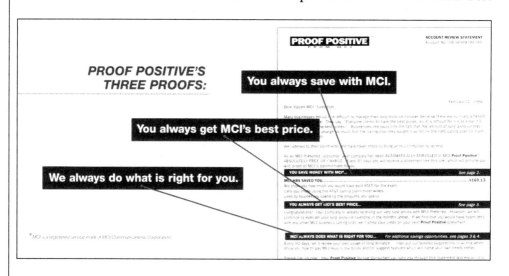

choice. First, to prove that MCI would always save the customer money over AT&T, it introduced a quarterly statement showing all of the customer's actual calls and what they cost using MCI versus what they would have cost using AT&T.

Second, to combat concerns that rapid new-product or program introductions would make the customer's current product or program obsolete, MCI ran the quarterly list of calls against all of its other products and programs. If it found that any other product or program could have saved the customer money during the period, it billed as though the customer had used that product or been a part of that program.

Third, to ensure that customers were using the right product or service for their businesses, the company promised continual follow-up to recommend new ways to save. It was simply the latest in a long line of marketing innovations, and for MCI the "proof positive" was in the 20 percent sales increase that followed.

The sales gains were remarkable in and of themselves, but Schmetterer considers Price to have produced an even greater achievement. Just as Friends & Family signaled a marketing breakthrough in the way consumers came to view long distance service as a brand-name product rather than as a mere commodity, so Price spawned a breakthrough in the way people thought about the business market by treating it as a mass market rather than a niche market. Up to this time advertising for business products generally meant running print ads in business journals such as *Fortune* or *Business Week*. Price, however, caused a paradigm shift by running a whole series of local and national television spots that appealed to the business customer as a consumer.

Other companies soon picked up on the success of this strategy and also started to advertise directly to business people in their homes via television. "We realized that everything we do in business also has an impact on people who are in the consumer market," Schmetterer said. "Again, creating ideas that were bigger than commodity ideas, we did a series of commercials about MCI customer reps doing whatever it takes to meet the needs of their business customers, emphasizing the spirit of the MCI employees."

In addition to speaking directly to the businessperson through mass media, Price's strategy also worked to support the efforts of MCI's business representation. Price understood that the business sales force is made up of young, vibrant people who form ongoing relationships with their customers. Unlike the telemarketer, whose job might only involve switching a customer's long-distance carrier, the business rep must call on clients and form relationships with them to land the big contracts that provide multilayered communications services encompassing both voice and data transmission. These relationships are nurtured through visits to customers to update and ensure that the services provided are meeting their needs.

Schmetterer likens Price's efforts to a battleground maneuver. "Tim understood brilliantly that advertising could provide this tremendous air cover for the ground war being fought by sales reps calling on businesses, trying to get in to talk to people, trying to sell competitively against others," he explained. "The air cover motivated the sales force and allowed them to say, 'Have you seen our advertising? Let me come in and tell you about what we are doing.'"

A New Way to Call Collect

As much as MCI enjoys surprising the public with completely new products, it is not above retooling old ones if it thinks they will sell. Such was the case with its 1993 introduction of 1-800-COLLECT. The company described the service as "America's least expensive way to call collect" and proceeded to capture a piece of the $3 billion market.

MCI marketed the service separately from other proprietary services because a key innovation was that customers could dial 1-800-COLLECT from any telephone regardless of the long-distance carrier. MCI left no doubt about its prime target, however, advertising that the new service offered savings of "up to 44 percent over AT&T."

In just two and a half years, MCI turned 1-800-COLLECT into a major profit center and in the process cornered more than 30

percent of the market. That achievement, perhaps more than any other of MCI's products, indicates how much the company's success is dependent on its marketing and advertising know-how. Whereas Friends & Family changed long-distance calling from a commodity to a brand, the 1-800-COLLECT campaign created a whole new product category where previously none existed. And to market the product, MCI turned to a sales force made up not of telemarketers or customer service reps, but of advertising spots. This product was wholly driven by advertising. As Schmetterer remarked, "You wouldn't know about it if it was not for advertising."

1-800-COLLECT revolutionized collect calling by giving consumers a choice in making collect calls in the U.S. and to more than 100 countries. Neither caller nor recipient needed to be an MCI customer.

Collect callers do not have a real incentive to care about price since they are not the ones paying for the call. MCI took this product category in which people had little reason to change providers and created a market based almost solely on attitude. For MCI to launch the service, no technological innovations were required, just a good gimmick.

"We decided to create a brand that would be based on emotions," remembered John Donoghue, senior vice president of consumer marketing. "The image is young, it's hip, it's a little bit edgy." Among the most popular commercials for the service was an ad featuring onetime David Letterman sidekick, Larry (Bud) Melman, dressed in a bumble bee outfit. It was Taylor's idea to position 1-800-COLLECT as the cool way to call because college students and other young people are the primary users of collect services. "It's an amazing thing: He [Taylor] took a hundred-year-old product, collect calling," observed Price, "and, on the basis of personality, turned it into a huge business."

That service produces more than $300 million in revenues annually. In addition to its profitability, 1-800-COLLECT is about as low maintenance as a product can be. There are no salespeople, and marketing is handled by just three MCI employees.

"It's the single biggest business success that I've ever witnessed," Schmetterer remarked with obvious admiration. "MCI created something where nothing existed, and in a very short time captured a huge market share—and did it really with creative thinking. This is a perfect example of everything MCI is about: creative thinking to develop the idea for COLLECT; creative thinking to execute that idea from a system standpoint; and then creative thinking to design compelling, interesting, fun advertising and, therefore, a personality for the product."

In true MCI fashion it took just forty days to conceive and plan 1-800-COLLECT to the point of being able to announce it to the media. And even more startling is that the operators in its service centers were successfully trained in how to handle the calls only forty-eight hours before the product was introduced. MCI kept the project under wraps because it was afraid AT&T would get wind of 1-800-COLLECT if details of the service were provided any earlier.

British Telecom Acquires Interest in MCI

Success in the national arena did not prevent MCI from keeping an eye on the international market. In the spring of 1993 it announced a deal whereby British Telecom (BT), a company with which it originally partnered in 1984, agreed to invest $4.3 billion in MCI.

Under the terms of the arrangement BT, the fourth-largest carrier of international communications traffic (behind Nippon Telephone and Telegraph, AT&T, and Deutsche Telekom), acquired a 20 percent interest in MCI. Furthermore, the two companies agreed to pour an additional $1 billion into a new alliance designed to offer communications services to international corporations. BT controls about 75 percent of the equity in the joint venture and MCI holds the remainder. In operational terms the two are full and equal partners.

It was a deal, however, that almost never happened. MCI initially walked away from the bargaining table because management felt BT just wanted too much. "We just couldn't get the final mile," recounted Doug Maine, MCI's executive vice president and chief financial officer. "We thought the control provisions and the governing provisions were just unreasonable given the size of their investment."

MCI then began serious discussions with Deutsche Telekom and French Telecom, but ultimately decided to give BT another go. In a five-day marathon session the deal was finally hammered out. The two companies entered a global strategic alliance through a joint venture called Concert Communications to market services and provide integrated voice and data links on a common network platform. Unlike traditional international corporate-calling services, which were established on a country-by-country basis in a patchwork of capabilities, Concert services allowed seamless voice transmissions. In essence Concert offered a private network serving businesses around the world.

Companies responded favorably. Most liked dealing with one communications company for all their international calling needs rather than one for every country. (In the 1997 planned merger with WorldCom, BT's 20% stake in MCI was bought out. As of this printing, the future of the Concert alliance is still uncertain.)

networkMCI—MCI Looks to the Future

The year 1994 opened with a surprise for MCI's competitors. In January the company announced networkMCI, an unprecedented $20 billion effort to create new global services based on the convergence of computers, telephones, and on-line access.

Roberts described this six-year corporate transformation in the press as "the sum of all the plans and opportunities we have in the new emerging markets."

In the short term networkMCI would offer e-mail, fax service, Internet access, paging, and video conferencing in a format that businesspeople had asked for—bundled together in one integrated service. To get this service off the ground, thirty dedicated employees from a variety of disciplines gathered in an MCI war room crammed with computers and telecommunications gear. Working around the clock, they tested and debugged the networkMCI software, just as McGowan's teams had tested new products and services years before in similar war rooms before bringing them to market.

The network will let the company bypass the Baby Bells and eliminate hefty access charges that cost MCI about $6 billion annually. MCI is building a new infrastructure—including digital driveways to bring the information superhighway to individual users—based on SONET, a powerful and superior high-speed fiber optic technology. MCI expects to spend $2 billion to install the fiber optic cables and switching equipment that will allow it to leapfrog the local Bell systems.

Most of the remaining $18 billion investment will go to upgrade MCI's fiber optic network in preparation for the advanced services to come including video-on-demand and multimedia applications. MCI's new network will then be linked to the rest of the globe through its alliances with companies such as BT, Stentor in Canada, and Banamex-Accival in Mexico. MCI customers will be part of the first global, fiber optics-based information system.

When MCI said it would offer connections to other long-distance carriers and charge them less than the local Bell companies charged, it caught the attention of AT&T. A spokesperson said, "If MCI sometime in the future can offer access at less expense and with equal quality to that we get now from the Bell companies, we would be foolish not to take advantage of that." AT&T, no slouch when it comes to spotting an opportunity, may now be ready to buy from its longtime adversary if it will be to its advantage.

InternetMCI—I Have Seen the Future and It's Exhilarating

Once again demonstrating its quickness and agility, MCI got the jump on AT&T with its InternetMCI, which the company positioned as the centerpiece of its emerging cyberspace business. InternetMCI offers customers a portfolio of on-line services including software for easy access to Internet sites and a high-speed fiber optic hookup for businesses.

By the time AT&T debuted its Internet service in March 1996, MCI was already earning $100 million annually on InternetMCI, for which it charges customers $19.95 for twenty hours of access a month. In an effort perhaps to make up for its late entry into the game, AT&T offered customers free service for a year, a costly freebie for a competitor already playing catch-up. Just how costly was indicated by the company's disclosure in September 1996 that earnings for the third and fourth quarters would not meet expectations: A heavy investment in new businesses like Internet access was cited as one of the drags on AT&T profits.

The speed with which MCI was able to bring the product to market had more than a little to do with Vinton Cerf, senior vice president of data architecture. Cerf, a technology maverick who has worked for MCI on and off since 1982, agreed in February 1994 (after some gentle arm-twisting by Bert Roberts) to help get InternetMCI off the ground. And who better to take on this task than the man who helped design the original Internet some twenty-five years ago? Cerf's journey onto the information superhighway officially began in 1968 when, as a graduate student in computer science, he became part of a UCLA team that

MCI Profile

Vinton G. Cerf

Vinton Cerf is senior vice president of Internet architecture and engineering for MCI. Cerf is responsible for the development of MCI's Internet network, the world's fastest and largest of its kind. He oversees the design and development of the network architecture that will enable MCI to deliver a combination of data, information, voice, and video for businesses and consumers.

A legend in the computer industry, Vint Cerf is known as the "Father of the Internet." He helped develop the networking protocol on which the entire system is based. *Wired* magazine has described him as a "Net celeb," and he also has the distinction of being the only MCI executive ever included by *People* magazine in its annual list of the nation's twenty-five most intriguing individuals. (He was chosen in 1994.)

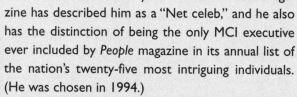

Most recently he was appointed by President Clinton to a committee to advise the High Performance Communications Office on plans for the "next generation" Internet.

This is Cerf's second association with MCI. He first joined the company in 1982, serving as vice president of MCI Digital Information Services where he was the chief engineer of MCI Mail.

In 1986 Cerf left MCI to join the Corporation for National Research Initiatives (CNRI) where he spearheaded research efforts on information infrastructure technologies. He served as vice president of CNRI until his return to MCI in 1993.

Cerf is a fellow of the Institute of Electrical and Electronic Engineers (IEEE), the Association for Computing (ACM), the American Association for the Advancement of Science, and the American Academy of Arts and Sciences.

developed the predecessor to the Internet, something called ARPAnet. Six years later he helped design the protocol that was the basic building block for Internet communications.

Cerf believes that technology will evolutionize, rather than revolutionize, our lives. "I don't take an apocalyptic view here," he explained. "I see technologies coming and they don't necessarily replace everything. They augment. Sometimes they reinforce. They often change ground rules and business models, costs, and things of that sort. But they don't necessarily wipe out previous things."

MCI's Internet product reinforces Cerf's beliefs. It offers important features to make life easier without foisting overly complicated technology on the user. With it customers can make secure transactions over the Internet without fear of having their credit-card numbers or other sensitive information stolen by electronic eavesdroppers and hackers. The software package also contains point-and-click technology that allows both consumers and business users to browse the Internet's World Wide Web over ordinary phone lines.

If, as Cerf predicts, there are more than 400 million people on 4.1 million networks by the end of the century, the Internet is a market that MCI could not afford to forgo. "I have seen the future and it's exhilarating," Cerf exclaimed.

A New Wrinkle in 1995

The frenzied pace Bill McGowan established during MCI's formative years seems to be in the company's genes. If anything, the pace accelerated in 1995 as MCI managed, yet again, to pursue aggressive expansion plans and introduce new services. But this time, MCI added a new wrinkle: It moved to become a global packager of content as well.

MCI announced in May 1995 that it intended to invest $2 billion in Rupert Murdoch's News Corporation, Ltd. The deal translated into a 10 percent equity stake with 13.5 percent voting rights in News Corporation and an option to increase that equity stake to 20 percent. Together the companies planned to deliver

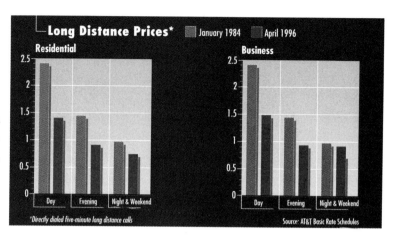

The price of a long-distance call has dropped dramatically during the last decade. In the residential market, the price of a five-minute phone call has declined 42 percent, from $2.40 to $1.40.

entertainment, information, home shopping, and other services to households and businesses worldwide. Each company was to invest $200 million in the joint venture initially, with News Corporation's investment coming in the form of assets.

The move was heralded in the press as bold, but few suggested that it was smart. "Down the road, MCI's strategy may seem masterful," Danny Brier, president of the consulting firm TeleChoice, told *USA Today*. "But right now, I don't see it." Among the chief criticisms was that MCI had gone Hollywood—that it was dazzled by Murdoch's colorful array of products from the Fox television network to the *New York Daily News*. The barbs didn't bother Roberts, however. He sees things very clearly.

"It's not that something is wrong with long distance," Roberts told *Forbes* magazine in explaining the diversification. "We're going to continue to grow that business. But other major communications industries are melding and moving into our territory."

Roberts also insisted that MCI has no intention of getting into the movie business, the newspaper business, or even the television business. What MCI does plan to do is maximize the use of its fiber optic infrastructure. The way to do that is by becoming involved in some of the content it is moving to its customers. But actually creating or producing in these businesses is not part of the game plan.

A Dizzying Array of Acquisitions

News Corporation wasn't MCI's only deal in 1995. The company also acquired Nationwide Cellular Service for $190 million in cash. At the time of the purchase, Nationwide was buying excess network capacity from GTE and other companies, then selling discounted cellular services to its customers. Nationwide brought with it 275,000 customers in ten cities including New York and Los Angeles. The move by MCI followed AT&T's acquisition of McCaw Cellular and Sprint's latest venture, an alliance with Tele-Communications, Comcast, and Cox Communications to deliver cellular services.

In June MCI announced that it planned to buy Chicago-based Darome Teleconferencing, the leading provider of financial and investor-relations conferencing, to further boost its conferencing services in its networkMCI Conferencing division. The deal would be completed for $32 million in cash and stock.

Earlier in the year MCI also purchased a 23.5 percent stake in Belize Telecommunications from BT for about $20 million in cash. Viewing Central and South America as a large, untapped market, MCI believes that the acquisition will enable it to better serve its business customers as it expands in the region.

In addition MCI continued efforts to get into the wireless paging market by joining forces with Paging Network and SkyTel

MCI's share price has reflected its success over the years, rising from $0.28 to nearly $33 (adjusted for share splits) by the end of 1996. That means that a $1,000 investment in MCI would have turned into $116,222 over the same period.

MCI Profile

Michael H. Salsbury

Michael Salsbury is executive vice president and general counsel of MCI. As the company's chief legal officer, Salsbury leads MCI's legal, regulatory, and legislative efforts including all patent filing, litigation, interface with federal and state regulatory officials, contract development, and support for both domestic and international sales as well as other legal matters.

Salsbury, who joined MCI in November of 1995, has been and continues to be a central figure in telecommunications law. He worked closely with Congress, the Senate, and the White House Administration to revise key sections of the Telecommunications Bill so that it more fairly served long distance carriers. Once the bill was signed into law on February 8, 1996, Salsbury worked closely with the federal regulatory team to develop the National Rules, which became the guidelines used by each state regulatory commission in determining terms and pricing for carriers providing local service in that state. Most recently Salsbury has lead MCI's legal efforts in access charge reform and the RBOCs' entry into long distance.

Salsbury, who was previously the managing partner of the Washington office of the law firm Jenner & Block, has a longstanding affiliation with MCI. He was MCI's primary counsel in proceedings involving the landmark AT&T consent decree and in competitive issues before the Federal Communications Commission and state regulatory bodies. Salsbury also supervised the regulatory and antitrust approvals for the BT–MCI alliance.

Corporation to provide nationwide messaging services under the name MCI Paging. Aimed at the growing mobile workforce, the service is being marketed through MCI's established national sales force. Immediately after the acquisition, MCI bundled the paging service with Friends & Family, making MCI the fastest-growing paging company in the United States, bringing in an estimated 60,000 customers per month.

As if all these acquisitions weren't dizzying enough, MCI boldly announced in March that it would go head-to-head with the regional Bell companies by offering local telephone service in ten U.S. cities including New York and Chicago. The company allocated a total of $500 million to initiate the service and expected to undercut the Baby Bells' prices by several percentage points. Baltimore was targeted for start-up because Maryland regulations permit local telephone-service competition; the service generated $38 million in sales in the first six months.

In the midst of these grand schemes, 1995 was also characterized as a time when MCI customers and would-be customers were offered a flurry of new deals. A 25 percent discount to all Friends & Family customers on any call placed in the United States came in January. In March MCI said it was diversifying into money-transaction services for banks, financial institutions, and retailers. It acquired the money-transaction technology from British Telecom, which had offered the service for almost twelve years. BT brought about two hundred clients of the service to MCI, including Corestates Financial Corporation and Budget Rent-A-Car Corporation.

But just as in past years, the array of strategic alliances and new products announced in 1995 had its down side. Investment analysts—and some investors—complained about what they viewed as a lack of a coherent deal-making strategy.

"The appeal of MCI was that it was such an easy company to understand when it was just a long-distance carrier," said William Deatherage, of investment banker Bear, Stearns & Company, an MCI corporate customer. "Now, it is becoming an increasingly complex corporation. There has been uncertainty as to how exactly MCI would diversify."

> By its steadfast refusal to accept change, to accommodate change, or to admit change, monopoly guaranteed change.
>
> – BILL MCGOWAN

In only twenty years, the long-distance market in the U.S. has transitioned from a monopoly to true competition. AT&T's effective 100 percent market share has dropped to 53 percent, and MCI is now a major player with nearly 18 percent of the market.

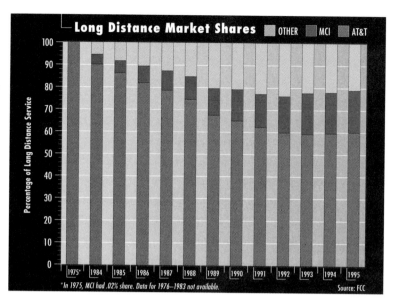

*In 1975, MCI had .02% share. Data for 1976–1983 not available. Source: FCC

To be sure, no telecommunications company seems to be going in as many different directions as MCI. Yet Roberts remained resolute: "We're increasingly frustrated by analysts who can't bring themselves to see the vision of this company and the way the industry is changing." Angela Dunlap believes that if MCI were to cubbyhole itself as just a long-distance provider, it would be missing out on a huge part of the market. Companies still want the long-distance service, but they are just as likely to want it packaged with computer and outsourcing services.

The Breakup of AT&T: Round Two

In 1995 AT&T made some startling announcements of its own. After fighting its divestiture in 1984 and making some significant acquisitions in the 1990s—among them McCaw Cellular, the country's largest wireless company, which it purchased for $11.5 billion—AT&T decided that bigger might not be better after all. AT&T began one of the biggest voluntary dismantlings in corporate history.

The surviving, if somewhat diminished, AT&T included its $53 billion long-distance and cellular phone business, while the Network Systems division (formerly Western Electric), which makes switching equipment, and Global Information Solutions, AT&T's computer division, were split into separate companies. The equipment business, including the renowned Bell Labs, was subsequently spun off as Lucent Technologies, Inc., and the computer business was slated to be reincarnated under the NCR Corporation banner by the end of 1996.

While this move seemed like an about-face, the downsizing was AT&T's stab at staying competitive in a market that it perceived would be deregulated imminently (as it was in short order—see the following chapter). Particularly in the case of Western Electric, the changing face of telecommunications had taken its toll. The local telephone companies, which had once been Western Electric's target market, would no longer buy equipment from their rival AT&T.

When the size of the layoffs accompanying the reorganization became apparent, AT&T chairman Robert Allen came in for some harsh criticism. Variously labeled "Scrooge," "corporate killer," and the "executioner," Allen faced a barrage of critics who were particularly incensed that the man responsible for cutting so many people from the payroll (the number of projected layoffs eventually rose to 40,000) received $5.2 million in his own pay envelope, not to mention stock options that boosted his total 1995 package to more than $10 million. Allen, who has

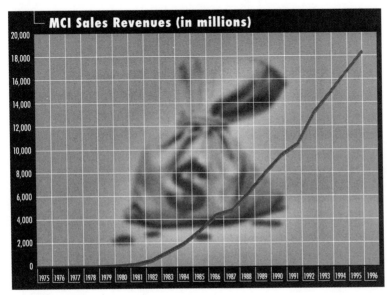

MCI's revenues have grown from $7 million in 1975 to more than $18 billion in 1996.

long had a reputation as a man of integrity and mild manners, explained his actions as painful but necessary to position the company for the future.

Defenders of Allen pointed to the unprecedented turbulence in the telecommunications industry and the increasing competition AT&T faced from all corners. Industry analysts were also quick to note that change was long overdue at the often too-little too-late communications giant.

As the decade passed the midpoint, AT&T had good reason to be worried. Statistics for 1995 showed that the company that once owned the long-distance telephone business lock, stock, and barrel had, in little more than two decades, seen its 100 percent market share dwindle to 53 percent. MCI had captured almost 18 percent of the long-distance traffic, U.S. Sprint had garnered 10 percent, and the remaining 19 percent had been taken by a host of smaller competitors.

Chapter 8

The Future Is Calling: 1997 & Beyond

> To succeed in anything, you have to take a few risks. Very little of consequence was ever created by those who always played it safe.
>
> —BILL MCGOWAN

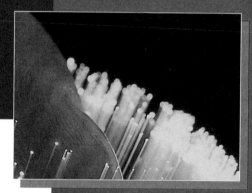

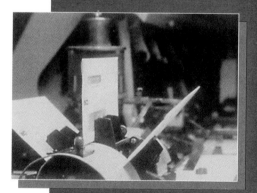

The Next Generation

On February 8, 1996, President Clinton signed into law the Telecommunications Act of 1996, ending government regulations that have existed since 1921. The barriers holding back companies from entering the local and long-distance markets, broadcasting, cable television, on-line, and wireless services were no more.

The law's signing reflected the administration's acknowledgment of what has long been clear: Telecommunications is an organic, quickly evolving industry that can no longer be constrained by archaic regulation. But dismantling the longstanding barriers is a complex procedure. So complex that after years of debate, Congress, the Clinton administration, and industry leaders agreed to let the FCC preside over the most serious and contentious issues such as how and when local carriers can offer long-distance service in their regions.

Even with many of the details still to be mapped out, the Telecommunications Act is a weighty piece of legislation, coming in at nearly 40,000 words, and the FCC itself has issued an additional list of one hundred new regulations. As the regulatory

Vice President Al Gore looks on as President Bill Clinton uses an electronic pen to sign the Telecommunications Reform Act, Thursday, February 8, 1996, at the Library of Congress in Washington, D.C.

pronouncements continue to multiply (the FCC released a more than five-hundred-page ruling in early August of 1996), it is becoming clear that the FCC intends to move aggressively to pry open the local telephone markets. Among the August pronouncements, the FCC ordered the Baby Bells to allow long-distance carriers to lease capacity on their telephone lines at a 22 percent discount to retail rates. The regional companies also must rent individual pieces of their network so competitors will not have to pay for more of the network than they need.

Access Charges Still a Major Issue

1970 logo

While the August actions were good news for MCI and the other long-distance carriers, they were still awaiting FCC action on the major issue of access charges. MCI now pays the regional Bell operating companies (RBOCs) almost $6 billion a year in access charges for service that costs them only about $1 billion to provide. The long-distance industry in aggregate pays about $22 billion for services that cost the RBOCs $3 billion to provide.

In a speech before the National Association of Regulatory and Utility Commissioners prior to the FCC's August ruling, Jerry Taylor explained why MCI believes that access charges must be cut to an at-cost level before the RBOCs are allowed to enter the long-distance business. "The Baby Bells have the highest operating cash flow margins of any industry in the country...larger by a factor of two than the long-distance industry," Taylor said. "[Local] access [is] the most lucrative line of business for the RBOCs, with margins about three and one-half times more than the long-distance industry." Citing the "unfair advantage" that the high level of access charges gives the RBOCs, he said that "competitors can't be saddled with inflated access charges that subsidize the RBOCs' entry into long distance if real competition is to be created in local markets and retained in long distance."

It will take years for the FCC to decide all the issues. That probably explains why, at the time of the President's signing, twenty-three states had already begun the cumbersome process of deregulating local-call monopolies. They know the FCC is cautious and likes to take its time. And, certainly, with legislation as monumental as this, the agency will be meticulous. Bell Atlantic, for example, petitioned the FCC to allow it to carry video services over its telephone lines in southern New Jersey. The FCC took two years to approve the petition; the state of New Jersey took just a few months. Whenever the final decisions are made, one thing is certain: The telecommunications industry, which is expected to gross $1 trillion in annual revenues by the year 2000, will be growing at a breakneck pace and changing with astonishing speed.

The Future Is Now

Even if President Clinton hadn't signed the Telecommunications Act, MCI would have made sure that 1996 was a memorable year in the industry. On January 25 it announced the purchase of what one reporter described as an "orbital parking spot," paying $682 million for a license to offer satellite-TV service across the United States. Shortly thereafter, MCI and partner Rupert Murdoch's News Corporation disclosed plans to start two direct-broadcast satellite (DBS) services. One, called American Sky Broadcasting, or ASkyB, will provide direct-to-home entertainment for consumers, while the other, SkyMCI, will provide data, information, and training services for businesses. In addition, just three days after its entry into the DBS arena, MCI announced a partnership for an on-line service with Microsoft Corporation to market Internet services on a mass scale.

1982 logo

1996 logo

These deals signal that the lines are fast blurring between telecommunications, computing, content, distribution, television, and software and hardware. The future of convergence, which Bill McGowan predicted more than a dozen years ago, is now. And MCI has long been operating on an assumption that the Telecommunications Act made official: No one can afford to operate only in one telecommunications market anymore.

Only after President Clinton signed the legislation, however, did MCI's grand plan (or non-plan, as the media was wont to characterize it) become clear to those beyond the corporation's

MCI Profile

Fred M. Briggs

As chief engineering officer, Fred M. Briggs is responsible for global engineering services, network systems engineering, network systems development, intelligent services platform, network architecture and advanced technology, data services engineering, and data architecture. Briggs oversees wireless engineering, strategic accounts engineering, the FAA LINCS program, and technical support services.

Briggs served on the Board of Directors for Concert, the joint venture between MCI and BT, as well as on the Board of Directors of InFlight Phone Corporation.

Briggs most recently served as senior vice president of network services engineering for MCI. Prior to that, he was senior vice president of network operations. Briggs also has served as vice president of data service engineering and vice president of data services for MCI International, where he was responsible for the international messaging and private line services. In 1983 he joined MCI to work in the company's international network planning organization.

Prior to joining MCI, Briggs held operations management and engineering positions with AT&T and Mobil Chemical Company.

inner circle. "It's no longer a question of how large a share of the $76 billion long-distance market you can get," UBS Securities, Inc. analyst Linda Meltzer told *Business Week*. "It's a question of how big a share of the $500 billion converged or integrated market you will get." To that end MCI is concentrating on selling as many services as it can. That's what all the partnerships, alliances, and new products are about. "We want to get as many hooks into each of our customers as possible," explained Roberts.

Shortly after the Telecommunications Act became law, MCI made an effort to package those hooks into an irresistible product that bundles services and equipment in one box and delivers it right to a customer's door. On April 29, 1996, it launched MCI One. Designed for consumers and small businesses, the product allows customers to create their own packages of telecom services by choosing from among cellular, paging, calling card, personal 800 numbers, long distance, Internet access, and home security offerings. The integrated services are then billed on a single statement and supported through a single customer service number.

Most impressive is MCI One's paging feature. When friends, family, or business associates call a customer's individual number, MCI's intelligent network goes to work to track down the customer by automatically trying a series of pre-programmed numbers such as office phone, home phone, and cellular phone. If the customer doesn't pick up, the call is then directed into voice mail and the customer is paged and alerted to the waiting voice message. Remarkably, all of this happens in a matter of seconds, unbeknownst to the caller.

Another feature of MCI One, the computing package, would seem to have great appeal for the technologically challenged: A personal computer already loaded with popular software is delivered and set up at the customer's home or office complete with on-site training. Before the delivery truck is out of the driveway, the customer's e-mail messages can be traveling through cyberspace.

> The lines are fast blurring between telecommunications, computing, content, distribution, television, and software and hardware.

MCI Profile

Michael J. Rowny

Michael J. Rowny is executive vice president of ventures and alliances for MCI Communications Corporation. In this capacity he is responsible for the operational oversight of MCI's alliances with British Telecommunications, Avantel, Stentor, and News Corp., and oversees MCI's strategic planning and corporate development.

Rowny was previously with MCI from 1983 to 1986, during which time he served as vice president and treasurer of the company and also as senior vice president of finance for MCI's domestic long distance business. He was a key player during the company's most challenging period of growth—the breakup of the Bell System and the beginning of equal access.

Prior to his return to MCI, Rowny was executive vice president and CFO of ICF Kaiser International, an environmental and engineering services company. His broad career in business and government included positions as chairman and CEO of the Ransohoff Company, CEO of Hermitage Holding Company, vice president for strategic planning and ventures of the Bendix Corporation, and deputy staff director of the White House during the Carter administration.

Rowny holds a bachelor of science degree from the Massachusetts Institute of Technology and a J.D. degree from Georgetown University Law Center. He is a member of the Young Presidents Organization and the District of Columbia Bar.

MCI is signaling its competitors that it is well positioned to greet the future even though the convergence of telecommunications, computer, and on-line services has yet to take a definable shape. "There's nobody in the world who doesn't believe that MCI, with its marketing and sales strength, can't pick off 15 percent of any market," Roberts told *Forbes* in a 1995 interview. "We could pick off that much of the shoe market, and we don't even make shoes."

Roberts's remark is a vintage MCIism—brash, cocky, and full of enthusiasm. It is with this same attitude that the telecommunications giant will meet the new century—even though neither Roberts nor anyone else knows exactly what the new millennium will bring except change. "In the year 2010, we will be looking at an industry that's nothing like the industry we see today," he said. "Now we have clear lines of demarcation...in the future they'll be melded into blended technologies."

> In some ways, the future is even more uncertain than it was in 1983 before the breakup of AT&T and the advent of equal access.

Changing the Way Children Learn

One area in which these technologies will have a huge impact—and where MCI hopes to play a part—is education. "You will find the use of a multimedia national information infrastructure in the schools and in the homes changing the way we do education in this country," Roberts said. This infrastructure will give far broader resources to teachers and students, ultimately making education more fun and more disciplined and allowing the country to achieve higher standards.

MCI, under Roberts's personal supervision, is attempting to directly affect education through CyberEd, a program that employs an eighteen-wheeler fully equipped with personal computers, Internet connectivity, and CD-ROMs as well as printing, faxing, and video-conferencing capabilities. All services are available free of charge to local educators, community leaders, and families. The truck travels around the country and visits disadvantaged areas in both urban and rural communities.

CyberEd was spawned by MCI's CyberRig, a mobile unit designed to show business people how easy it is to use the latest in

> Bert Roberts believes that by the year 2000, 50 percent of MCI's revenue sources will come from services and products that did not exist in 1996.

communications and information technology. The project is funded by a coalition of nonprofit, corporate, and private foundation partners, all brought together through Roberts's efforts. Among them are Microsoft Corporation, Corning Incorporated, and the Milken Family Foundation.

In October 1996 MCI took its interest in cyber-education a step further by joining with Sylvan Learning Systems, Inc., a Columbia, Maryland-based provider of educational services to families, schools, and industry, to form an international distribution network for adult professional education services. Called the Caliber Learning Network, the new venture will utilize state-of-the-art technology to deliver undergraduate- and graduate-level university courses, Continuing Education programs, and customized corporate-training-and-development programs throughout the United States and around the world. Centers equipped with satellite-based video-conferencing capabilities will be located in all major U.S. cities, and home study will be provided through the Internet. About fifty Caliber centers are expected to open in 1997, with further expansion both at home and abroad envisioned for 1998.

Besides providing the communications and computer-networking services, MCI is also investing $10 million for a minority stake in the venture. Having expressed his belief that technology will change the way we educate, Bert Roberts is making sure that MCI is an active advocate of that change.

Visions of the Changing Market

Roberts goes so far as to predict that the information superhighway will similarly change MCI. He believes that by the year 2000, 50 percent of MCI's revenue sources will come from services and products that did not exist in 1996. That is a remarkable statement in light of the fact that long-distance service still accounted for 90 percent of MCI's business at the end of 1995 with about 40 to 45 percent of that coming from the residential market and the rest from the business sector.

But large long-distance companies like MCI will be most able to take advantage of the market free-for-all spawned by the Telecommunications Act. "The long-distance carriers have the marketing skills, the brand awareness, and the technological know-how to put together packages of local, long-distance, data, and video services," T. Mark Maybell, managing director of Merrill Lynch & Company's telecom investment banking group, told *Business Week*. "Long term, that puts them in better shape to become twenty-first century 'supercarriers'—single sources for all communications services."

That doesn't mean, however, that there is clear sailing ahead for MCI. At the beginning the upstart had only one competitor—a behemoth to be sure, but still only one. Today, MCI faces challengers at every turn whether in the local and long-distance arenas or in wireless communications, satellite broadcasting, and international telecommunications. Moreover, the domestic regulatory issues are still not completely resolved, and the increased competition from the RBOCs can't be dismissed lightly. MCI may be gaining access to local markets but at the same time its viable competition in long distance and other product areas is expanding.

In some ways, the future is even more uncertain than it was in 1983 before the breakup of AT&T and the advent of equal access. Wall Street analysts are quick to point out that the risks are many. And Jerry Taylor agrees. "The biggest challenge we have is figuring out how to get into local business, and it's terrifically complex. It's complex technically," he said. "It's complex legally."

The big plus for MCI is that it has grown up in a competitive marketplace, whereas the RBOCs still have to learn what it means to compete for customers. And although the company recognizes that this is a critical time, a time in which every part of the business will be tested, it also strongly feels that it has the edge on its competitors. "After all, it's not like this isn't something we haven't done before. We've gone up against monopolies very successfully," Tim Price wryly noted. "We've got 500 competitors today; we'll have 507 tomorrow."

MCI's relationship to its customers, therefore, is also evolving. No longer is MCI primarily concerned with simply acquiring new customers, but rather it is concerned with building value for existing customers by expanding the breadth of service it offers them.

The New Global Frontier—Emerging Key Players

The telecommunications industry is constantly evolving. The most recent major events include a pact signed by seventy countries in the World Trade Organization to open telecommunications markets and the planned opening up of phone markets in the European Union in 1998. This dissapearance of borders, along with the fact that about half the world's population lives two hours from a telephone and has never placed a phone call, makes this time one of huge growth potential and opportunity in telecommunications. The follwing list covers the key players and their latest maneuvers in this dynamic, highly energized industry.

NTT (Nippon Telegraph and Telephone) — $79 Billion

- National Japanese Phone Company; government-owned.
- Monopolistic Japanese carrier with divestiture expected before 2000.
- 1996 government decision to maintain as holding company with three subsidiaries.
- *Key International Effort*: None as of 1997. All international service provided by one of three Japanese telecommunications companies—KDD, ITJ, and IDC.

AT&T (American Telephone and Telegraph) — $52 Billion

- Number one U.S. provider, with about 53 percent market share, number two worldwide behind NTT.
- *Key International Effort*: WorldPartners, a loose federation of sixteen phone companies including Japan's KDD, Singapore Telecom, and Europe's Unisource Consortium.

British Telecommunications plc — $24 Billion

- BT shares 90 percent of the U.K. market with Mercury Communications.
- *Key International Effort*: Aside from primary U.S.–Europe alliance, other investments include those in New Zealand's Clear and France's Cegetel.
- Since broken partnership plans with MCI, now looking for a new entry into U.S.

Deutsche Telekom — $39 Billion

- German national phone company; gradually privatizing.
- *Key International Effort*: Global One Alliance with France Telecom and Sprint.

MCI: Failure Is Not an Option

France Telecom — $30 Billion

- French national phone company, first public shares offered in 1996.
- *Key International Effort*: Global One Alliance with Deutsche Telekom and Sprint.

MCI WorldCom — $28 Billion

- New U.S.-based company formed from the merger between MCI and WorldCom, now sharing 25 percent of long-distance market.
- Also substantial player in U.S. local and Internet markets.
- *Key International Effort*: Heavy Internet presence in Europe, fiber optic lines in major cities for voice and data transmission, and substantial presence in the international calling business.

Telecom Italia — $20 Billion

- State-owned Italian phone company, through holding company STET (privatization set for 1997).
- *Key International Effort*: Joint venture with Bouygues in France.

Telefonica de España SA — $16 Billion

- National phone company in Spain; competition to be introduced beginning December 1998.
- Joint venture with BT/MCI (Telefonica Panamerica MCI) to provide service to the $36-billion Latin American market, which is expected to double by the year 2000.
- *Key International Effort*: Major presence in Latin America through Tisa, its international holding company. In talks to partner with U.S. company to tie North and South American markets.

Sprint — $14 Billion

- Number three carrier in the U.S. with about 10 percent market share.
- *Key International Effort*: Global One Alliance with Deutsche Telekom and France Telecom.

Cable and Wireless plc — $8 Billion

- London-based; provides telecommunications services primarily to small and midsize companies.
- *Key International Effort*: Fifty-seven percent stake in Hong Kong Telecom (link to China market) as well as stakes in Mercury Communications (U.K.), IDC (Japan), Optus Communications (Australia) and Paktel (Pakistan).

> To succeed in anything, you have to take a few risks. Very little of consequence was ever created by those who always played it safe.
>
> – BILL MCGOWAN

Bob Schmetterer, who considers February 8, 1996, as "the first date of the next century," believes that the companies that understand that they are now in a much broader business than the one they've always been in will seize the opportunity and be big, big players. Companies will need to be inventive and creative, and MCI is already at the top of the list of companies that have shown these qualities. He also points out that MCI already has a reputation in the business markets as a proven provider of varied telecommunications products, and says its corporate customers "can't wait to turn it all over to them."

Integrating Services

Transmission

In its effort to chart its course in this new world of fast-paced changes in telecommunications capabilities and services, MCI management has come to see customer service, marketing, billing, and transmission as four core parts in a single, integrated structure that serves as its bridge to the consumer. Its telemarketing and customer service business, for example, may prove to provide a large share of future revenue. As the world's largest telemarketer, it sells not only its own services but the services of many other companies as well. And what sets MCI's operations apart is, not surprisingly, its innovative technology. Its customer service platforms are flexible enough to allow the same center that sells and services Friends & Family to take orders for individual 800 numbers and answer Microsoft Network customer service requests. This segment, which already contributes hundreds of millions of dollars to annual revenue, is the fastest-growing part of the company's business, according to Price.

Marketing

Transmission is the core segment of MCI's business and includes everything from long-distance phone calls and faxing to sophisticated data transmissions. It involves MCI's intelligent network, which according to Price is a completely digital network that has "intelligence built above the switch level," making it possible to distinguish relationships. Calls do not have to be billed in a formulaic way, therefore, in a manner such as X number of miles times Y number of minutes equals Z amount of money. Instead, as in the Friends & Family program, for example, the network can determine that phone numbers are in a customer's calling circle and so provide a discounted rate accurately accounted for.

This sophisticated transmission component, in turn, feeds the billing system, which has to be robust enough to handle two hundred million transactions every day. Price emphasizes that in a world of electronic commerce, volume is key, so success depends on having the capability and the intelligence to correctly transmit and bill an enormous volume of transactions.

Customer service

The company believes that every business in America someday will intersect with MCI on one of these four components, and once a connection is made, a large proportion of the customers will choose to expand their business to one or more of the other components. For example, a company seeking MCI's telemarketing services might then decide to utilize its billing component. Microsoft came to MCI looking for a way to handle customer service for the biggest software launch in history—Windows 95. Today, MCI handles all of Microsoft's phone services and the company represents MCI's biggest outsourcing agreement. "Microsoft wanted a customer service solution, and we got its transport business as well," Price noted. "Now the company wants marketing and billing services, too."

Billing

By providing these services to other companies, MCI is moving toward its goal of utilizing its infrastructure to expand its

British Telecom: Ally or Competitor?

The telecommunications industry was officially born in the U.K. in 1879 when the British Post Office won an exclusive right to operate the telegraph network. Once private companies began to develop telephone systems, the British courts protected the post office monopoly by empowering it to license private telephone companies, to collect royalties in the amount of 10 percent of revenues, and to own and operate its own telephone system. A private company called the National Telephone Company was the Post Office's main competitor. When this company's license expired in 1911, however, the Post Office staged a takeover of National Telephone and became an outright telephone monopoly.

In 1981 the government took away telecommunications services from the Post Office and placed them in the hands of a new company—British Telecommunications Corporation. The government also allowed for competition, most prominently from Mercury Communications, the second largest telecommunications venture in the U.K. The government soon began to apply pressure for BT to privatize. After the Telecommunications Act of 1984 was passed, BT went public in a big way with a publicity campaign estimated at $20 million. (The BT offering was the largest single stock offering in the history of Great Britain.) The Telecommunications Act created the Office of Telecommunications (OFTEL) as a regulating body and set the rates BT could charge, which were indexed to inflation.

Throughout the 1980s BT was intent on expansion into the global marketplace. In North America it purchased control of the Canadian phone equipment company Mitel in 1986, the American electronic-mail services company ITT Dialcom in 1986, and a 20 percent stake in U.S.-based McCaw Cellular in 1989.

In 1990 new competitors emerged including U.S. cable companies (many owned by RBOCs in the States), France Telecom, cellular-telephone service companies, and other large corporations. In 1993 BT came into its own when the government sold almost all of its remaining stock in the company for $7.4 billion.

In 1994 BT purchased a 20-percent stake in MCI for $4.3 billion. BT also had a victory in the courts when Mercury lost a bid to force OFTEL to alter its regulations regarding the way BT charged rivals to use its network.

BT continued to maneuver successfully as it sold its share in McCaw Cellular to AT&T for approximately $1.8 billion in 1994 and established a presence on mainland Europe by investing $1.7 billion in the new French phone-service company Cegetel in 1996. Now, with the recent dismantling of the proposed Concert alliance with MCI, BT will be searching for other ways to expand outside of its European borders—preferably into the U.S. Many believe that AT&T will be the next potential partner.

business. It is not getting into a new venture, as it has appeared to some observers, but rather simply building on its existing base.

MCI's core business, transmission, as well as the content that is being transmitted, are evolving rapidly. MCI's relationship to its customers, therefore, is also evolving. No longer is MCI primarily concerned with simply acquiring new customers, but rather it is concerned with building value for existing customers by expanding the breadth of service it offers them. According to Taylor such expanded service may mean that MCI reaps increased revenues when, for example, a customer who was spending twenty-five dollars a month is now spending a hundred dollars a month, and the customer reaps huge benefits in increased power of communication and convenience—a true win-win situation. Taylor explained, "it's not just about giving us more revenue and profit. It's also about us becoming much more valuable to that customer." As MCI continues to position itself as a supercarrier that is a premier supplier of both communications and customer service, it will be further removed from its one-time image as the cheap choice but will still remain price competitive.

By the 1990s, as industry price wars took their toll on all the major long-distance companies by seriously eroding profit margins, MCI had already begun moving toward competing on integrated services. Starting with Friends & Family, it began an intense advertising campaign that promoted integrated services. It also came up with special integrated services for large accounts including customer-service call centers, processing of credit-card charges for some retailers and banks, and assistance in setting up their own internal communications networks. Now, MCI One takes the concept of integrated services another giant step forward.

In the global arena MCI is looking to expand in developing countries that are making telecommunications their number one infrastructure priority. China, for example, is putting up phone lines at the rate of ten million a year, which represents half the new phone lines in the world.

MCI and WorldCom: An Alliance for a New Age

If the MCI-WorldCom alliance is approved, MCI will be undergoing a tremendous transformation on a number of different levels. Most importantly, MCI's business will expand far outside of long-distance services, now reaching into virtually all aspects of telecommunications. This evolution has been inspired by a combination of opportunity and necessity. Closed markets of the past—including U.S. local phone and foreign long-distance—have begun to open at an astonishing rate. And new markets, including that of Internet access, are being created where previously none existed.

Of course, the transition will not be easy. Even with the advantages of MCI's current market share, sound financial footing, and powerful partners, the monopolistic forces—whether RBOCs in the U.S. or foreign giants protecting their turf—once again must be fought in the courts as well as in the marketplace. These companies are still using delaying tactics to slow progress, just as AT&T did in the days of equal access. But competition has been good for consumers, and consumers know this for a fact (which they did not before equal access), and it is they who will push out the monopolies. It is the customer's demand for competition that is transforming the telecommunications industry.

MCI-WorldCom, the new company formed by the merger of WorldCom and MCI, will be able to serve customers in widely disparate ways. The company will be able to offer unified services such as long distance, local, and Internet services—all combined into one bundled package. This consolidation is also propelling MCI into entirely new fields of systems integration, such as networks, company intranets, and electronic commerce.

The use of teams, a hallmark of MCI management structure, will be a critical aspect of the successful transition to the WorldCom alliance. MCI has been extremely successful with this problem-solving technique—using one team to get one job done, and then assembling a different group for a different task. Partnership, perseverance, and flexibility will be the key to MCI WorldCom's success in the emerging competitive global marketplace. It is the MCI spirit—the willingness to do things that need to get done—that ultimately may be the most important legacy MCI will bring to its new alliance.

Financial Markets Still Count

The ability to raise capital will continue to be central to MCI's efforts to grow and expand its reach. However, capital is no longer a scarce resource for MCI. The company has proven itself worthy of receiving capital whether it's from joint-venture equity deals, bank or vendor financing, or the capital markets.

"Today, of course, we are a single-A," CFO Doug Maine noted, referring to MCI's credit rating by Moody's Investors Service. This status as a company that can easily attract capital will keep it in good stead as it tries to stay out in front of the competition and adapt to changing technologies.

Although some skeptics think the imminent convergence of the telecommunications industry that MCI has been touting is more rhetoric than reality, few argue with the company's position for taking advantage of convergence when it finally does come about. "The market has expectations that MCI can increase its double-digit growth rate from its core business and grow new businesses internally without any degree of...operating losses that would dilute consolidated earnings growth," reported a telecommunications analyst for Lehman Brothers, echoing the thoughts of many market insiders.

Doing What Works

Although MCI still maintains the demeanor of a feisty upstart, it is a vastly different company than it was in Bill McGowan's heyday. In that era the company pushed the bankruptcy envelope on more than one occasion. Scarred veterans of those years sweated out every monthly profit-and-loss statement, never knowing if the end was coming.

But the end did not come. With a powerful entrepreneurial spirit that McGowan described as combining moxie, chutzpah, hubris, and lunacy, MCI thrived. And as the years passed, the company grew ever stronger, absorbing many valuable lessons along the way, which are explicated in the following pages. From strategy and finance through marketing, operations, and human resources, MCI's executives learned what works and what does not, and they were quick to make adjustments, even change direction when they had to. That characteristic adaptability remains one of the company's greatest strengths, one that can carry it into the twenty-first century and keep it in the forefront of its industry.

MCI has faced such a staggering array of problems, particularly in the early years, that it's sometimes hard to understand how the company survived. But the explanation from those who made the journey and helped construct one of America's great telecommunications empires would no doubt be a simple one: they did it because they had to. Failure was not an option!

Opposite: Bill McGowan and company.

Afterword

It would be foolish to claim any kind of closure to a story as mercurial as MCI's, or to an industry with the frenetic pace of telecommunications. But even by MCI's standards, the pace of the last few months has been breathtaking. Since the completion of the main text of this book only a few months ago, MCI has gone through multiple merger announcements. One history-making alliance has been announced, then adjusted, and finally abandoned for another even more dramatic partnership. As always, MCI has been right in the thick of the excitement.

As these pages go to print, it seems increasingly likely that MCI will be merged with another upstart telephone company, WorldCom, Inc. As with so many aspects of MCI's story, this step will be a record-breaking event. At a total value of $37 billion, the proposed merger will be the largest in history, with the closest second being the Bell Atlantic-NYNEX merger worth $26 billion.

It has been a tumultuous trip to this latest turning point. Prior to WorldCom's bid, MCI's most significant suitor had been British Telecom. In November of 1996, BT offered to buy MCI for $37 per share, in a deal that was hailed as groundbreaking for its international scope. But as subsequent months passed, both sides became increasingly uneasy about the inherent problems involved in the combination of two such different corporate cultures. (One analyst characterized BT as "Masterpiece Theater to MCI's MTV.") Although BT has come a long way from its monopolistic past, it is nevertheless a very different company than the freewheeling MCI.

As these culture-clash concerns became more prominent, MCI's earnings took an unexpected dip in the second quarter of 1997 because of expenses related to its move into the local phone business. BT reacted by cutting its offer dramatically, from $39 to $32 per share, or from a total of about $25 billion to $18 billion. This was exactly the opening that WorldCom needed and it made an unsolicited all-stock offer for MCI at $41.50 per share. This offer was followed by a $40 per share cash offer from GTE, and suddenly MCI was the prize object of a bidding war once again. After several rounds of negotiations and revised offers, WorldCom stunned the investment community with a final offer of $51 per share, or approximately $37 billion (all in WorldCom stock). They sweetened the offer even further by offering the position of chairman to Bert Roberts, as well as several senior positions to other top management guns at MCI. This last offer was a generous one, and on November 10, 1997, MCI accepted this final bid.

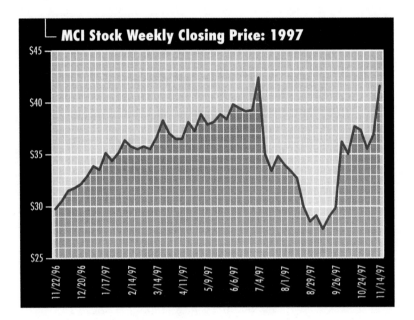

MCI Stock Weekly Closing Price: 1997

The alliance behind MCI-Worldcom, as the company will be called, looks promising. The synergies between the two companies seem ideal, with WorldCom's local phone business being the perfect compliment to MCI's heavyweight position in long-distance. This combination will make MCI-Worldcom the first truly comprehensive communications company, with the ability to offer its customers real "one-stop shopping." WorldCom also brings to the table the largest presence on the Internet, meaning that the new company can offer all of the major telecommunications services (with the exception of cellular) to its business customers. WorldCom-MCI will have 25 percent of the long-distance market, local connections in about 100 U.S. cities, and the most significant Internet connections in the country. Having all of these functions under one roof will be an exceptional advantage, to say the least.

But the similarities in style between the two companies may be even more important than any business synergies. From this perspective, MCI and WorldCom seem like a match made in heaven. Many aspects of WorldCom's story echo those of MCI's struggle. The company was born in a coffee shop brainstorming session between Bernard Ebbers, its CEO, and three friends in 1983. They incorporated it as LDDS (Long-Distance Discount Service) in Mississippi, as a reseller of long-distance minutes. After a difficult first two years, the company acquired its first long-distance customer, the University of Mississippi. "Bernie" Ebbers, who took over as full-time president in 1985, took that modest start and never looked back. Through a series of leapfrogging acquisitions—usually of larger companies—WorldCom became the fourth largest carrier in the country by 1992.

Many analysts have attributed the success of the deal to the easy relationship between Ebbers and Bert Roberts. Ebbers personal style certainly fits in with MCI's irreverent culture—he's often characterized as the "cowboy" of the industry, and can usually be found walking around WorldCom's headquarters in Jackson, Mississippi, in faded jeans. He spends most of his free time on his cattle and soybean ranch, riding his tractor for relaxation. And as with so many of MCI's key leaders, Ebbers has been praised as having exactly the kind of fresh approach that the telecommunications industry needs. A former high school coach, he is most credited for his ability to "think out of the box." His fans point to his ability to instinctively understand the new competitive landscape of communications, perhaps because he has never really been a part of its regulated past.

With Ebbers as president and CEO, and Roberts as chairman, the media has taken to calling the new company the "Bert and Bernie" show, a nickname that neither of the two men seem to mind. (Ebbers is determined to demonstrate Robert's continuing authority, and refers to him as "bossman".)

Who knows what lies ahead? That question, of course, is impossible to answer. This deal still faces the ever-vigilant eye of the Justice Department, and MCI's experience with British Telecom proved that no merger is final until the papers are signed. As for the years ahead, the future of telecommunications will no doubt be filled with more change that will be as exciting as it will be chaotic. Or as Bert Roberts said recently: "when you're in a telecom revolution, sometimes things don't happen in an orderly manner."

Timelines

Part 2

A Visual History of MCI and the Telecommunications Industry

Telecommunication, destined to be one of the world's largest industries in the 21st century, originated with the invention of the telegraph by Samuel F.B. Morse in the 1830s. Dozens of telegraph companies cropped up in the 1840s and early 1850s, and in 1856 a group of them came together to form Western Union, which completed the first transcontinental connection in 1861.

The telegraph not only established a new industry but set the pattern for future developments. For the first time, information could be dispatched instantaneously by mechanical means. In the past, it was transmitted by human and animal effort, mostly along routes established for human transport. Now, information went wherever wires could be established. Further, the telegraph put other means of communication out of business. The Pony Express, only months in service, was eliminated in 1861. "The pony was fast," wrote one historian, "but he could not compete with lightning."

In time, the telegraph was displaced by the telephone. Alexander Graham Bell patented his technology in 1876, and commercial service began the following year. Soon the rival technologies of telegraph and telephone clashed.

During this period, most business leaders considered themselves wedded to their technology, and the divisions between industries were quite clear. Western Union was a telegraph company and determined to defend its technology against interlopers. At that point, Western Union could have purchased the Bell interests for $100,000 but rejected the opportunity. It finally entered the field in force, employing Thomas A. Edison to develop technologies alternate to Bell's. But it was too late to displace Bell Telephone. The conflict was resolved in 1879 when Western Union agreed to abandon its foray into telephones in exchange for a 20 percent interest in Bell's rental receipts.

Telegraph would remain for another century, but as telephone connections grew, the telegraph became a supplemental means of communication and eventually all but disappeared. Unable to come to terms with technological change, Western Union faded.

Led by Theodore Vail, Bell transformed into American Telephone & Telegraph (AT&T), which dominated long distance while licensing other companies to provide local service. Vail proved an outstanding manager, and under his leadership AT&T became the primary telecommunications company, gobbling up local companies, creating a superb research operation in Bell Laboratories, and acquiring Western Electric, which became its manufacturing arm. AT&T customers had to use Western Electric equipment and local operating companies had to cooperate with AT&T if they wanted to offer services outside of their areas. AT&T even threatened to acquire Western Union and dominate all aspects of telecommunications.

In 1913, to avoid antitrust prosecution, Vail agreed to divest AT&T of its Western Union shares and purchase no additional independent telephone companies without FCC permission. AT&T further agreed not to enter related industries. The "Kingsbury Commitment," named after the AT&T executive who negotiated the arrangement, remained the basis for telecommunications policy for well over half a century. With this, AT&T became a "regulated public utility," accepting government regulation in return for the preservation of its dominant position.

The Kingsbury Commitment both helped and hurt AT&T. As part of this arrangement, AT&T offered low-cost local service that was subsidized, in part, by higher costs for long distance. This meant that long-distance customers were charged higher rates than economies might indicate, while local consumers received a better deal than they knew. It also meant that AT&T functioned in a uniform environment with fixed rules that enabled it to make long-term plans and improve services gradually.

Under this rubric, Americans enjoyed the best and most consistent telephone service in the world. But it also mandated that the many inventions that flowed out of Bell Labs, including transistors and other micro devices, software, solar-energy collectors, and numerous computer improvements, could not be produced and marketed. AT&T might well have become a major player in motion pictures and radio were it not for the constraints imposed in 1913. The Kingsbury Commitment did not halt technological progress, but for a while it blocked the emergence of the telecommunications industry while preserving the status of telephony.

AT&T was aware of technological challenges to its position. One of those came from wireless, invented by Guglielmo Marconi in 1895. At first, the technology seemed supplemental to the telegraph, of particular use for contacting ships at sea. Radio Corporation of America, formed in 1919 from what originally was American Marconi, seemed poised to challenge Western Union and perhaps AT&T. But led by David Sarnoff, RCA instead entered into broadcasting, a completely separate industry, and the telecommunications possibilities of wireless were downplayed for the time being.

Radio's emergence increased federal regulation of the industry. Telecommunications was to be governed by the Communications Act of 1934, which created the Federal Communications Commission to oversee the industry.

AT&T's quasi-monopoly started to crack in the 1960s in ways that would transform telecommunications forever. Carterfone developed a system to connect telephones to mobile radio transmitters. When AT&T threatened to disconnect customers who used the device, Carterfone sued and won FCC approval in 1968, ending AT&T's monopoly on telephonic equipment.

Microwave Communications Inc. presented a major challenge to AT&T. In 1963, MCI asked FCC permission to furnish private-line service between Chicago and St. Louis. MCI's plan was to use microwave transmission, which was less expensive to create, operate, and maintain. Cost considerations, plus the absence of the need to subsidize local calls, enabled MCI to offer business customers substantially lower rates than those available from AT&T. Another company, Datran, asked for similar permission, and in the end, the FCC granted both requests.

Then the Justice Department filed an antitrust suit against AT&T in 1974. The suit was dropped in 1982 when AT&T agreed to divest itself of its local operating companies, then known as the "Baby Bells" and now also referred to as regional holding companies (RHCs) or regional Bell operating companies (RBOCs). This meant that AT&T was now freed from the constraints imposed by the Kingsbury Commitment. Thus the new era began, its outcome still unknown.

The major players in long distance were AT&T and MCI. Sprint was the third major player, entering the field in the 1980s. They were joined by the wireless companies, which attempted to create a mobile telephonic system using cellular technology. Some of the Baby Bells were also quite interested in expansion but were restrained from entering long distance under the terms of the decree by which they came into existence.

Wireless is one of the most glamorous and promising technologies in telecommunications. Its customer base is growing at the rate of 40 percent a year, with an expected 100 million users by 2004. Wireless is usually dated to 1946, when mobile-telephone service was attempted in St. Louis, but it really began with Marconi and citizen-band radio. In 1970, AT&T demonstrated an improved system in Chicago. Then in 1982, the FCC accepted applications for licenses in 30 cities with plans to permit two competing systems in each. Dozens of companies filed applications, and soon after, hundred of bidders appeared when 60 other cities were opened.

Established companies soon entered the wireless field. Sprint acquired Centel in 1993. Southwestern Bell purchased the cellular business of Metromedia and later changed its name to SBC Communications, as though to indicate its new directions. Bell South acquired the cellular business of Mobile Communications with MCI. More followed. The other regional holding companies—Bell Atlantic, NYNEX, U.S. West, Ameritech, and Pacific Telesis—also tried to find ways to enter into strategic alliances and mergers to expand their businesses.

McCaw Cellular Communications was the most important of the newcomers. Craig McCaw, who took over a family-controlled cable television company after his father's death in 1969, turned first to paging and then to acquisitions, the most important being MCI's national cellular rights, for which he paid $120 million, and LIN Broadcasting, which had valuable cellular properties and cost $3.4 billion. He cut back on his debt by selling 22 percent of his company to British Telecom, enabling that company to enter the American market.

In this way, McCaw Cellular became the industry's largest player. McCaw's ambition in the early 1990s was to displace AT&T as the premier provider of services. It appeared unlikely, but considering that cellular is inexpensive, more flexible, and easier to use than wired telephones, it did not seem so wild a dream. Indeed, McCaw suggested his biggest competition might well come from cable television companies planning to enter telephony.

Unlike Western Union more than a century before, AT&T understood the situation and reacted swiftly. When McCaw partner British Telecom paid $4.3 billion for a 20 percent stake in MCI, the money was to be used to finance a venture into cellular. AT&T astonished the industry by announcing it would purchase McCaw for $12.6 billion. Thus AT&T became not only the largest wired telecommunications company but the leading player in wireless as well.

The arrival of new technologies in the 1990s spurred the creation of scores of start-up companies, which complicated the industries and blurred the divisions between telecommunications, cable television, and other industries. U.S. West formed an alliance with Time Warner, from which a union of cable television and telecommunications may spring. In response, Bell Atlantic entered into an aborted bid to acquired Tele-Communications. Likewise, a venture between Southwestern Bell and Cox Communications, a cable firm, failed to come about.

Pacific Telesis took a different route, divesting itself of its cellular operations and then announcing a union with SBC Communications to form the second largest telecommunications company in the nation. Pacific Telesis also indicated an intention to concentrate on the emerging personal communications services (PCS) market.

PCS, which uses a higher frequency band than cellular, initially served only local needs but in time could compete with cellular. Nextel, organized in 1987 as Fleet Call, provided what it called "specialized mobile radio" to several large communities and was one of the most important of the start-ups.

Clearly this was an industry in flux, and the outlines of what it will become by the turn of the century are by no means clear. The situation became even more fluid when, in early 1996, President Clinton signed a new telecommunications bill permitting the regional Bell companies to offer long-distance services outside of their areas.

New technologies spawn even newer ones. It would not be surprising to see an expansion of the Internet into telephony. McCaw, the most original telecommunications entrepreneur on the scene, had a plan of his own. After selling his McCaw Communications interest, he joined Microsoft's Bill Gates to organize Teledesic, which plans to launch and operate 840 satellites at a cost of $9 billion. This network would transmit telephone calls anywhere on earth. It would be akin to a telephonic version of the Internet and might bear a relationship to the Internet, similar to the one that once existed between the telegraph and the telephone.

The concept was McCaw's. More than a century ago, Alexander Graham Bell dreamed of such a system, and Thomas Alva Edison toyed with the notion. Princeton physics professor Gerard O'Neil organized Geostar, which was based on a similar notion. Motorola has plans for Iridium, a global telephone service based on 66 satellites, and other companies have similar concepts. This is not unusual; many major inventions have come out of intense competition between inventors struck by the same thought at the same time. If Teledesic, Iridium, or another newcomer becomes a reality, it could mean that every computer will connect with millions of other computers, that each person could have his or her own portable telephone and an individual number, and be able to contact anyone else in a flash—a vision that bears some resemblance to that of the individuals who marveled at the telegraph when it displaced the Pony Express in 1861.

The History of Telecommunications 1837–1997

1837 Telegraph invented by Samuel F.B. Morse

1856 Western Union formed

1840s Telegraph companies emerge

1861 Growing telegraph industry puts the Pony Express out of business

1844 First public telegram sent

1861 Western Union completes first transcontinental connection

1830s   1840s   1850s   1860s

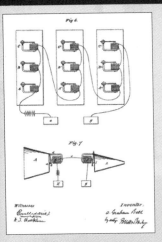

1876 Alexander Graham Bell patents telephone technology

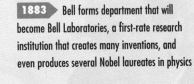

1883 Bell forms department that will become Bell Laboratories, a first-rate research institution that creates many inventions, and even produces several Nobel laureates in physics

1885 American Bell Telephone's first use of the name American Telephone & Telegraph

1878 Commercial telephone service begins

1893–94 Alexander Bell's original patents expire, triggering competition from hundreds of local companies

late 1870s Western Union hires Thomas Edison to develop alternative technologies to Bell

1879 Western Union agrees to abandon foray into telephones in exchange for a 20% interest in Bell's rental receipts

1895 Guglielmo Marconi invents wireless "radio" communication

1870s 1880s 1890s

The History of Telecommunications 1837–1997

1907 A group of New York bankers led by J.P. Morgan takes over AT&T; they institute plan of buying up regional phone companies

1921 Congress passes the Graham-Willis Act, which deems AT&T a natural monopoly—a business characterized by economies of scale such that one firm can supply an entire market in an economically viable manner

1934 Radio's emergence increases federal regulation; the Communications Act of 1934 creates the Federal Communications Commission

1913 AT&T agrees to divest itself of Western Union shares to avoid an antitrust suit, and further agrees not to enter related industries. With this, AT&T becomes a regulated monopoly.

1900s 1910s 1920s 1930s 1940s

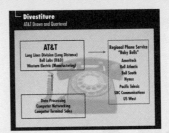

1984 AT&T breakup

Early 1960s Entrepreneur Tom Carter develops a device, dubbed the "Carterfone," that connects telephones to mobile radio transmitters

1971 FCC approves competition in special markets

Early 1990s Wireless growth; the FCC auctions licenses, and AT&T buys McCaw Cellular

1968 AT&T threatens to disconnect customers who use the Carterfone. Carter sues and wins FCC approval

1996 Telecommunications Reform Act is signed, tearing down barriers within the industry, and allowing telephone, cable and others to compete in each other's markets.

1968 Microwave Communications incorporates

| 1950s | 1960s | 1970s | 1980s | 1990s |

MCI Communications Timeline 1963–1997

⟶ LEGAL ⟶ FINANCIAL ⟶ TECHNOLOGICAL ⟶ CORPORATE ⟶ COMMERCIAL

MCI Born

October 1963 Microwave Communications, Inc. is organized in Joliet, Illinois, by five mobile-radio salesmen led by John D. (Jack) Goeken. The group aims to build a private point-to-point microwave communications system along Route 66 between Chicago and St. Louis for small businesses and truckers.

McGowan Arrives

February 1968 Bill McGowan and Jack Goeken meet for the first time.

Cash Flow Crisis

February 1967 Operating on a shoestring, Microwave Communications moves from one financial crisis to another, relying on additional contributions from the original partners, credit cards and occasional revenue checks. A "selective" offering of stock to current shareholders raises $30,000—enough to keep up the fight at the FCC, but still not enough to build the network.

Microwave Communications of America, Inc. is Incorporated

August 1968 McGowan invests $35,000 for 50 shares of MCI stock and pays the start-up costs for the new company. Ownership of MICOM is divided into quarters: 25% each to McGowan, Goeken, the original investors, and future additional investors.

1963
November
JFK assassinated

1965
March
Vietnam War escalates

1968
April
Robert Kennedy assassinated

November
Martin Luther King assassinated

Jerry Taylor Joins

August 1969 The 26-year-old Taylor agrees to join MCI at a significant pay cut. Under his leadership in the marketing division, MCI develops some of its most famous ads, spoofing AT&T by telling callers to "reach out and touch someone...but do it for up to 30, 40, even 50 percent less." After playing major leadership roles in the development of Execunet, as well as Friends & Family, he goes on to become president and chief operating officer.

MCI Wins New Permits

August 1969 In what is known as the MCI Decision, the FCC commissioners vote by a margin of 4-3 to approve the construction applications of the upstart company. It is the beginning of the end for AT&T's almost 70-year monopoly, and earns Jack Goeken the label, "Giant Killer."

MCI looks to Satellites

1971 MCI, Lockheed Aircraft Corporation, and Communications Satellite Corporation (COMSAT) form a joint venture to build a $168 million satellite system. Two years later, MCI, badly needing cash to expand its microwave network, sells its interest to IBM.

FCC Approves Competition

May 1971 The FCC approves widespread competition in private-line and specialized common carrier markets by issuing the Specialized Common Carrier Decision. The ruling ratifies the regulatory policies initiated by the MCI Decision and establishes the first guidelines for competition.

1969

July
First moon walk

August
Woodstock Music Festival takes place

1970

May
Kent State protesters killed

MCI Communications Timeline 1963–1997

> LEGAL   > FINANCIAL   > TECHNOLOGICAL   > CORPORATE   > COMMERCIAL

Chicago–St. Louis Service Begins

January 1972 — MCI links commercial customers between Chicago and St. Louis, with a string of microwave towers spanning Illinois at 30-mile intervals. Construction costs for this first segment of the MCI network are estimated at $4 million, nearly all financed by Raytheon, which has also agreed to a moratorium on payments until MCI began service. Irwin Hirsh memorializes the historic opening of the MCI system with: "I'll be damned, it actually works!"

Initial Public Offering

June 1972 — In the second-largest IPO of a development-stage company in U.S. history, MCI raises more than $33 million by issuing 3.3 million shares at $10 per share. With financing in place, MCI is now prepared to build its network.

$64,000,000.00
$64 Million Credit Line

June 1972 — Stan Scheinman, MCI's CFO, designs an innovative package of bank loans, supplier guarantees, and public stock issuance. In a sign of banker confidence in MCI's future, a group of investment banks led by First National Bank of Chicago signs a $64 million line-of-credit agreement.

Roberts Joins MCI

October 1972 — Bert C. Roberts, Jr., who would later be Chairman and CEO of MCI, joins the company.

1971

August
Nixon orders wage and price freeze

1972

September
Munich Olympic Massacre

Crews Speed New Network

1972

MCI Offers Specialized Services

January 1973 MCI becomes the first telecommunications company to market specialized services to the public. The company introduced 4K Plus, a new private-line service at much more competitive rates.

Antitrust Suit Against AT&T

1973 After efforts at direct negotiation fail, MCI first turns to Congress and eventually the courts in an effort to be granted interconnections for all services. In December 1973, a judge rules in MCI's favor and orders AT&T to provide the interconnections immediately.

Antitrust Suit Against AT&T

March 1974 MCI files a civil suit charging AT&T and all of its 24 companies with violation of the Sherman Antitrust Act. These suits mark the beginning of a bitter battle with AT&T that will last an entire decade.

1973

May
Watergate hearings begin

October
OPEC oil embargo

1974

August
Nixon resigns

MCI Communications Timeline 1963–1997

▶ LEGAL   ▶ FINANCIAL   ▶ TECHNOLOGICAL   ▶ CORPORATE   ▶ COMMERCIAL

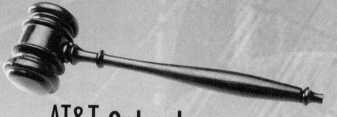

AT&T Ordered To Connect

April 1974 After a hearing, a U.S. District Court issued an injunction ordering AT&T to provide MCI with necessary interconnections. On appeal, a U.S. District Court of Appeals vacated the injunction, saying that the District Court had acted prematurely before the FCC had the opportunity to consider the matter. Days later, in a ringing affirmation of its pro-competition policy, the FCC gives AT&T ten days to provide MCI with the same interconnections and facilities it supplies to its own Long Lines Division.

Satellite Venture Sold

July 1974 IBM and the Communications Satellite Company (COMSAT) sign a memorandum of understanding to buy out the interests of MCI and Lockheed Aircraft in the CML Satellite Corporation. MCI receives $2.5 million from the sale.

Goeken Resigns

July 1974 Jack Goeken, president of MCI and the "sparkplug" of its original five founders, tenders his resignation as an officer and director of the corporation. He cites differences of opinion with other corporate officers and his lack of input into the decision-making process as his reasons for leaving.

Execunet Debuts

August 1974 MCI launches Execunet, a new private-line network tailored to meet small-business communications needs. Although the results in the original test market, Washington D.C., are disappointing, Execunet proves a success in Texas. By March 1975, Execunet's revenue surpasses its expenses.

1974   **1975**

April
Vietnam saga ends with fall of Saigon

FCC Halts New Service

July 1975 Responding to complaints from AT&T, the FCC orders MCI to stop offering its new Execunet service. MCI files suit with the U.S. Court of Appeals and receives a stay of the order, marking the start of three years of litigation over Execunet.

Dallas One Switching Begins

September 1975 Dallas One, a larger, faster, and more technically advanced computer switch designed to handle the growing number of Execunet calls, begins operation, providing long-distance service to 15 cities. By 1979, Dallas One handles 40,000 calls a day, and before it was retired in 1981, the switch processed an estimated $31 million in annual revenue for MCI.

MCI Forced TO OFFER STOCK

November 1975 Desperate for cash to expand, MCI raises $8.5 million in an equity offering which combines common stock and warrants. McGowan calls the deal "a giveaway," because of the rock-bottom share price at $2/share.

$8,500,000

THE ENTREPRENEUR AND THE MANAGER

December 1975 V. Orville Wright is elected MCI's president and chief operating officer. Forced by the lending banks to find an experienced manager skilled in running a large corporation, McGowan turns to Wright, who had been an executive at IBM, RCA, and Xerox.

1975　　　　1976

July
Bicentennial celebration

MCI Communications Timeline 1963–1997

> LEGAL > FINANCIAL > TECHNOLOGICAL > CORPORATE > COMMERCIAL

BANKS
Defer Payments

May 1976 — MCI's lending banks agree to defer interest payments on the condition that MCI hire a more experienced chief financial officer. Wayne English is hired shortly thereafter.

MCI ANNOUNCES
Its First Profit

December 1976

FCC AGAIN REJECTS EXECUNET

May 1976 — After a court-ordered review of Execunet, the FCC formally rejects the service once again, claiming it was not the private-line service MCI had been authorized to sell. The service continues, pending further litigation, but the negative publicity sends MCI common stock plummeting.

Loan Schedule STRETCHED

April 1977 — MCI and its lending banks sign a revised credit agreement which stretches its debt payment schedule.

1976 | **1977**

January
Roots *draws 80 million viewers*

APPEALS COURT
BACKS MCI

July 1977 The U.S. Court of Appeals vote unanimously to allow MCI to offer Execunet without restrictions.

DOUG MAINE
SIGNS ON

May 1978 Maine joins MCI, later joking that he's held every job in the entire billing department. He becomes executive vice president and chief financial officer in 1992, and directs a balance sheet restructuring that gives the company nearly $100 million in pretax savings.

AT&T FORCED
To Interconnect

April 1978 Following a Court of Appeals ruling which specifically orders AT&T to make Execunet interconnections available, AT&T agrees to provide them but at triple their current price. MCI executives appeal to the FCC, which agrees with them and urges AT&T to avoid further court costs and negotiate a reasonable settlement.

WEST COAST
LINKS ADDED

June 1978 MCI purchases key microwave routes to California from Western Tele-Communications, Inc. (WTCI) for $6.5 million. The system includes routes between Phoenix, Tucson, San Diego, and Los Angeles, as well as between Denver and San Francisco, giving MCI two routes from the Midwest to the West Coast.

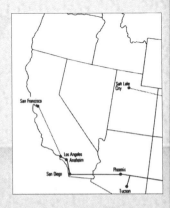

1977

December
U.S. has admitted over 165,000 boat people

1978

July
First test-tube baby born

MCI Communications Timeline 1963–1997

▶ LEGAL ▶ FINANCIAL ▶ TECHNOLOGICAL ▶ CORPORATE ▶ COMMERCIAL

NEW AT&T RATES NEGOTIATED

September 1978 Negotiations begin between AT&T and MCI to determine charges under the Exchange Network Facilities for Interstate Access (ENFIA) agreement. Roberts suggests that specialized carriers pay access charges tied to their level of market share; as long-distance revenues rose, so would the percentage they pay to the Bell operating companies. AT&T agrees.

Offering Raises $28.6 MILLION

December 1978 MCI completes a public offering of 1,229,850 shares of new convertible preferred stock paying an annual dividend of $2.64 per share. Priced at $25 per share, the stock sale nets $28.6 million.

STOCK SALE NET $69.5 MILLION

September 1979 MCI completes a public offering of 4.95 million shares of senior cumulative convertible preferred stock, realizing $69.5 million. MCI earmarks half of the amount for network expansion, with the rest going to buy back additional outstanding warrants, to pay off part of the bank credit agreement, and to add extra working capital.

NEW ANTITRUST SUIT Against AT&T

April 1979 MCI files a second antitrust suit — this one for $3 billion in damages — against AT&T, for its actions from 1975 to 1978. By striking a deal with a law firm to cut its fee in exchange for a percentage of damage rewards, MCI could do what none of AT&T's earlier rivals had been able to do: afford a drawn-out legal battle.

1978

November
Jonestown mass suicide

1979

January
OPEC begins drastic oil price increases

1. (MUSIC UNDER THROUGHOUT) MAN: Have you been talking to our son long distance again? WOMAN: (NODS AND WHIMPERS)
2. MAN: Did he tell you how much he loves you? WOMAN: (NODS AND WHIMPERS)
3. MAN: Did he tell you how well he's doing in school? WOMAN: (NODS AND WHIMPERS AND CRIES)
4. MAN: All those things are wonderful.
5. What on earth are you crying for?
6. WOMAN: Have you seen our long distance bill?
7. ANNCR: (VO) If your long distance bills are too much, call MCI.
8. Sure, reach out and touch someone. Just do it for up to 30, 40, even 50% less.

MCI The nation's long distance phone company 1-800-624-2222

RESIDENTIAL MARKETING BEGINS

March 1980 Calling itself "the nation's long-distance phone company," MCI begins serving residential customers, with Denver as its first test city. MCI's television and print ads emphasize potential savings of up to 50 percent over AT&T. One ad reads: "You haven't been talking too much, you've just been paying too much."

RECORD ANTITRUST AWARD

$1,800,000,000

June 1980 In the largest monetary award in American legal history, a jury in Chicago, Illinois, awards MCI $600 million in its 1975 antitrust suit against AT&T. Under antitrust statutes, the award is automatically tripled to an unprecedented $1.8 billion. The case is held up in the appeals process for the next five years.

$50 MILLION DEBENTURE ISSUE

$50,000,000

July 1980 MCI completes a public offering of subordinated debentures, raising $50.5 million. The offering, overseen by Shearson Loeb Rhoades, marks MCI's first venture into the public long-term debt market.

$51.4 Million OFFERING

$51,400,000

October 1980 MCI sells 3,630,000 shares of $1.84 convertible preferred stock at $15 per share. Managed by Shearson Loeb Rhoades, the offering nets $51.4 million.

1979
November — Iran seizes American embassy

1980
August — Poland approves unions, Solidarity founded
November — Ronald Reagan elected
December — Prime rate tops 20 percent

MCI Communications Timeline 1963–1997

▶ LEGAL ▶ FINANCIAL ▶ TECHNOLOGICAL ▶ CORPORATE ▶ COMMERCIAL

DEBT VS. EQUITY

April 1981 ▶ MCI sells $125 million worth of subordinated debentures, using the net proceeds of $102.1 million to pay down its debt and continue network construction. This is the first MCI offering handled by Drexel Burnham Lambert, which co-managed it with Shearson Loeb Rhoades. Bill McGowan meets with Michael Milken, the head of Drexel's high-yield bond department, for the first time.

$102,000,000

DEBT VS. EQUITY

August 1981 ▶ Drexel and MCI team up again to sell another $100 million of debentures, which MCI used for network construction and to retire preferred stock.

$100,000,000

BONDS NET $246 MILLION

May 1982 ▶ MCI continues the string of successful offerings it began in fiscal 1982 by netting $246 million in a public offering of convertible subordinated debentures. Proceeds from the sale pay for MCI's acquisition of Western Union International, as well as for network construction and new equipment. MCI's bond offerings greatly improve its financial position, which investors applaud. MCI's stock is the most actively traded on the NASDAQ in 1981.

$246,000,000

WESTERN UNION INTERNATIONAL PURCHASE

December 1981 ▶ MCI purchases Western Union International (WUI) from the Xerox Corporation for $185 million. Melding the disparate corporate cultures is a challenge, but the purchase of WUI's overseas base gives MCI a springboard into the international arena.

1981

January — Iran releases U.S. hostages
March — Reagan wounded by assassin
April — First space shuttle flight
July — Charles and Diana wed
July — First artificial heart implant

Quarter-Billion-Dollar BOND Sale

September 1982 After years of struggling for access to capital, MCI suddenly finds itself one of the hottest companies in the bond market. Encouraged by the success of its May offering, MCI sells $250 million of bonds.

$250,000,000

WATTS and FIBER OPTICS ARRIVE

January 1983 MCI orders more than 150,000 miles of fiber optic cable — the largest order ever placed.

New Satellite PURCHASE

February 1983 In what was the largest purchase of satellite capacity in telecommunications history, MCI agrees to purchase 24 satellite transponders, or radio relays, from Hughes Communications. The 48,000 new circuits help the company to handle the new subscribers it expects with nationwide equal access.

RATING UPGRADE and $400 Million Raised

March 1983 In its largest public securities offering yet, MCI sells $400 million of convertible bonds. With the new offering complete, MCI had raised more than $850 million in one year of debt securities issues, more than twice as much as the previous 10 years combined. That same month, Moody's rating service upgrades MCI to Baa3 — investment grade status.

$400,000,000

1982   1983

March
Reagan proposes Star Wars defense program

September
First African American Miss America crowned

MCI Communications Timeline 1963–1997

▶ LEGAL　　▶ FINANCIAL　　▶ TECHNOLOGICAL　　▶ CORPORATE　　▶ COMMERCIAL

One Billion Dollar OFFERING

July 1983 ▶ MCI raises $1 billion in a single unprecedented offering, and in the largest non-utility bond issuance in corporate history.

U.S.-Canada Service Begins

April 1983 ▶ MCI opens direct-dial service from the U.S. to Canada, announcing that its rates for most U.S.-Canada calls are 30 percent lower than AT&T's.

MCI Mail debuts

September 1983 ▶ CEO Bill McGowan introduces MCI Mail at a nationwide video news conference, saying, "We live in an instant information economy, and ours is the postal system designed for it." Although the new service puts MCI on the cutting edge of the information age, MCI Mail does not catch on with consumers, and is cut back to a bare-bones service by 1985.

AT&T makes the break

January 1, 1984 ▶ After nearly a century of dominating American telephone communications, the Bell System is finally broken apart in a court-ordered divestiture. AT&T divests itself of two-thirds of its assets, including 22 local Bell operating companies.

1983

October
U.S. invades Grenada

1984

December
Fatal gas leak in Bhopal

EQUAL ACCESS ARRIVES

January 1984 In the "Great Long Distance Telephone Election," every telephone customer in the country votes to choose a long-distance carrier, as directed under the 1982 Equal Access Consent Decree. The combined phone companies spend an estimated $500 million on a media blitz.

Tim Price joins

Month 1984 He goes on to play a pivotal role in creating brand identities for MCI's products, with success stories like Proof Positive and Friends & Family. He is later named president and chief operating officer of the MCI Telecommunications Corporation subsidiary.

MCI realigns

Interconnection rates rise / PRICE WARS begin

January 1985 MCI reorganizes its telecommunications division into seven regional units, each approximating the territory covered by one of the new regional "Baby Bells."

May 1984 In the wake of AT&T's divestiture, the rates MCI pays to the regional Bell companies for local telephone interconnections rise sharply. AT&T also increases the pressure by lowering its rates across the board by 6%, which MCI matches. These rate cuts mark the opening salvos in one of the most intense price wars in U.S. business history.

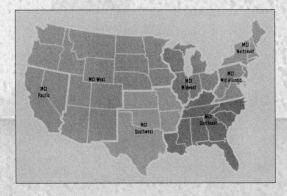

1984 **1985**

May
New Coke is introduced

MCI Communications Timeline 1963–1997

▷ LEGAL   ▷ FINANCIAL   ▷ TECHNOLOGICAL   ▷ CORPORATE   ▷ COMMERCIAL

IBM and MCI swap assets

June 1985 — MCI stuns industry experts by announcing an agreement to acquire Satellite Business Systems (SBS) from IBM in exchange for shares equal to a 16% interest in MCI. MCI's stock shoots up 33%. The arrangement proves to be largely unsuccessful, and MCI eventually buys back its stock from IBM.

Out-of-court SETTLEMENT

November 1985 — After more than a decade of antitrust litigation, MCI and AT&T settle two major lawsuits out of court. MCI collects an estimated $247 million. For the first time in its history, MCI is out of the litigation business—but only temporarily.

$575 million BOND ISSUE

April 1986 — Taking advantage of low interest rates, MCI offers 10 percent subordinated debentures and raises $560 million.

$575,000,000

Airsignal sold

July 1986 — MCI sells Airsignal, its paging and cellular phone business, to McCaw Cellular Communications for $116 million. The cellular market once seemed a perfect fit, but the company decides to concentrate on what it knows best: long-distance phone service.

1985   **1986**

December
Reagan signs Gramm-Rudman bill

January
Challenger *disaster*

April
Nuclear accident at Chernobyl

McGowan suffers
HEART ATTACK

December 1986 ▸ While vacationing in Virginia Beach, CEO Bill McGowan suffers a severe heart attack. Three days later, John deButts, the former AT&T chairman who led the company's efforts to block competition and preserve its monopoly, dies. The two events symbolize the end of an era of sweeping changes in telecommunications history.

800 service begins

March 1987 ▸ MCI jumps into the multibillion-dollar, toll-free long-distance market by introducing MCI 800 Service. By the end of 1988, MCI has captured nearly 10% of the domestic 800-number market.

1-800...

FIBER OPTICS
in place

January 1987 ▸ MCI's new coast-to-coast fiber optic network begins operations four years after the purchase of the more than 150,000 miles of cable needed to build it. The new cable dramatically increases MCI's carrying capacity.

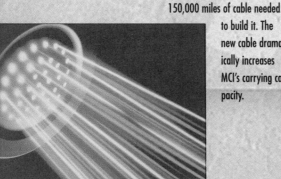

Operators
PLUG IN

January 1988 ▸ MCI takes another step toward full service by establishing operator services. Throughout 1988, MCI phases in other services, including assistance with collect calls, third-party billing, and international calls.

1986 | **1987**

January
Special committee begins to investigate Iran-Contra affair

October
Stock market crash

MCI Communications Timeline 1963–1997

▭▭▷ LEGAL   ▭▭▷ FINANCIAL   ▭▭▷ TECHNOLOGICAL   ▭▭▷ CORPORATE   ▭▭▷ COMMERCIAL

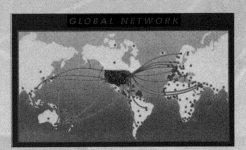

FIBER OPTIC NETWORK GROWS

July 1988 ▷ MCI announces construction of a new fiber optic route between Houston and Los Angeles. The 1,630-mile link will give the company 2 transcontinental fiber links.

GLOBCOM purchased

May 1988 ▷ MCI buys RCA Global Communications, Inc., an international carrier of data and telex communications, for $160 million from General Electric as part of its strategy to gain access to additional international voice markets. MCI International is now able to claim that it is the only broad "full service" international carrier.

MCI Fax unveiled

November 1988 ▷ The nation's first dedicated network for facsimile messages is unveiled at a New York press conference. McGowan describes it as part of MCI's plan to become a "one-stop source for global communications services."

1-900 service launched

September 1988 ▷ MCI announces the first nationwide 900 service to allow two-way conversations. MCI's new technology also allows users to interact through either voice command or telephone keys.

1988 | **1989**

May — Soviets quit Afghanistan
May — Reagan visits Soviet Union
November — George Bush elected
March — Exxon Valdez oil spill
June — Tiananmen Square massacre

Advertising
WAR BEGINS

May 1989 — MCI fires the first salvo in what would begin a bitter advertising war between the big three phone companies. The brawl eventually leads to AT&T's "Put It in Writing" campaign, an admission that MCI is a serious competitor.

MCI Acquires Telecom USA
Market Share Breakdown (Pre-Merger)

- Telecom USA 1.4%
- Metromedia-ITT 1.0%
- Others 7.4%
- MCI 12.9%
- US Sprint 8.6%
- AT&T 68.7%

Source: Dataquest Estimates

Telecom*USA acquired

May 1990 — In the largest merger in telecommunications history, MCI purchases Telecom*USA — the nation's fourth largest long-distance company — for $1.25 billion.

New Service to MEXICO

January 1990 — Telefonos de Mexico and MCI agree to provide direct voice service between the U.S. and Mexico. Later that year, MCI establishes the first fiber optic cable connection with Mexico.

1991–1994
FRIENDS AND FAMILY debuts

March 1991 — McGowan introduces the new Friends & Family calling plan, calling it "the biggest news in long distance since MCI itself." By November 1992, Friends & Family is used by eight million households, representing a 17% market share in the residential market.

1989
- **November** — East Germany opens its borders
- **December** — U.S. invades Panama

1990
- **February** — Nelson Mandela is freed
- **Augustr** — Iraq invades Kuwait
- **October** — Germany is reunited

MCI Communications Timeline 1963–1997

> LEGAL > FINANCIAL > TECHNOLOGICAL > CORPORATE > COMMERCIAL

Digital CONVERSION complete

December 1991 After taking a $550 million write-off on its analog system, MCI completes conversion of its entire network to digital transmission.

ROBERTS takes the reins

December 1991 Bert Roberts, MCI's president and chief operating officer, is named to the additional post of chief executive officer. McGowan steps down after more than two decades as CEO, but retains his position as chairman of the board of directors.

FAA awards contract

March 1992 MCI wins a $1 billion Federal Aviation Administration contract to upgrade the nation's air traffic control communications systems.

PROOF POSITIVE launched

Month 1992 MCI launches "Proof Positive," a product that offers businesses three proofs that MCI is their best choice.

1991 | **1992**

January Operation Desert Storm begins
June Boris Yeltsin elected president of Russia
August Kremlin coup attempt
September U.S.S.R. dissolved

McGowan Dies

June 1992 MCI founder Bill McGowan, who had received a heart transplant in the spring of 1987, collapses and dies of heart failure. Business and government leaders, friends and foes alike pay tribute. John Worthington, MCI general counsel and a confidant for 25 years, says simply, "MCI is his monument."

1-800 RULING

August 1992 The FCC rules in favor of "portability" of 800 numbers. The ruling strips AT&T of another marketing advantage, by allowing businesses to switch long-distance carriers without changing their 800 number.

1-800-COLLECT debuts

May 1993 Billed as the "least expensive way to call collect," MCI introduces an old product that has been retooled in an innovative new way.

Alliance with BT

June 1993 MCI and British Telecommunications (BT) agree in principle to form a worldwide global alliance.

1992
- April — L.A. riots
- November — Bill Clinton elected president

1993
- February — World Trade Center bombing

MCI Communications Timeline 1963–1997

▶ LEGAL ▶ FINANCIAL ▶ TECHNOLOGICAL ▶ CORPORATE ▶ COMMERCIAL

NETWORK MCI

January 1994 ▶ MCI announces an unprecedented $20 billion strategy to create new global services based on the convergence of computers, telephones, and entertainment.

Local phone service in the works

October 1994 ▶ The MCI Metro subsidiary files applications to provide service in several states, asking regulators to permit MCI to enter the local telephone market.

NEWS CORP. Alliance

May 1995 ▶ MCI stuns the business world by announcing plans to invest $2 billion in Rupert Murdoch's News Corp. The companies plan to deliver entertainment, information, and other services to households and businesses worldwide.

InternetMCI UNVEILED

November 1994 ▶ MCI announces InternetMCI, a portfolio of services including an electronic mall, software for easy access to databases on the Internet, and a high-speed fiber optic hook-up for businesses.

1994

April
Free elections in South Africa

1995

September
O.J. Simpson stands trial for murder

NATIONWIDE ACQUISITION

September 1995 MCI acquires Nationwide Cellular to establish immediate presence in the $21 billion cellular market.

MICROSOFT ALLIANCE

January 1996 MCI forms major product development marketing alliance with Microsoft to address the on-line, Internet, and networking markets.

MCI wins satellite license

January 1996 MCI pays $682 million for a license to offer satellite-TV service across the United States.

INTERNET 2000

March 1996 MCI unveils Internet 2000, with plans to build Internet business to $2 billion by the year 2000 by tripling the capacity of its U.S. network.

TELECOMMUNNICATIONS BARRIERS REMOVED

February 1996 President Clinton signs the Telecommunications Act of 1996, ending regulations that had existed since 1921.

1995

April
Oklahoma City Bombing

November
Yitzhak Rabin assassinated

1996

MCI Communications Timeline 1963–1997

> LEGAL   > FINANCIAL   > TECHNOLOGICAL   > CORPORATE   > COMMERCIAL

MCI ONE

April 1996 — MCI introduces MCI One, a single-source service that will allow customers to combine their long-distance, cellular, paging, Internet access and e-mail services, and be billed on a single statement.

SKY MCI

April 1996 — MCI and News Corp. announce the formation of American Sky Broadcasting (ASkyB) and SkyMCI to provide multimedia satellite services.

MCI-WORLDCOM MERGER

November 1997 — WorldCom announces planned acquisition of MCI for $37 billion, creating the first communications company to span the local, long-distance, and Internet markets.

CONCERT INTERNETPLUS

June 1996 — Concert InternetPlus, the first high-speed, high reliability global Internet backbone, is introduced by MCI and BT. The new network will increase the overall capacity of the Internet by an estimated 30 percent.

1996 | 1997

November — Clinton re-elected

Living Lessons

Part 3

- Management Strategies
- Marketing and Advertising Strategies
- Financial Strategies

The lessons that can be learned from the strategies used by MCI are broad and far-reaching. MCI began as a small entity among giants. The people moving MCI forward by necessity had to be inventive, persistent, daring, cooperative, combative, outlandish, impulsive, calculating, demanding, flexible, gutsy, shrewd, clever, responsive, and caring.

The lessons they have to teach are as applicable for the owners of small, mom-and-pop upstarts as they are for executives at large, global conglomerates. They include such subjects as how to create a potent workforce, how to boost employee morale and effectiveness, how to take advantage of the government regulatory environment, how to encourage growth, profitability, and brand recognition, how to compete with companies larger than your own, how to deal with capitalizing your company and creating worth, how to navigate in troubled financial waters, how and when to go public, how to be creative in financing, and how to be both self-reliant and interdependent in your field or industry.

Each lesson includes advice founded in MCI's storied experience and simple and effective action steps that can be taken immediately to further the goals of the lesson. By reading and integrating these lessons into your business, you should be able to avoid many of the mistakes and misfortunes that are the pitfalls of companies as they endeavor to sustain growth and prosperity.

Chapter 9

Management Strategies

The simple truth is that early on we didn't know what we were doing half the time. And the other half of the time we were dead wrong.
— BILL MCGOWAN

Hire Good People

Hiring self-motivated, loyal, passionate people involves more than evaluating qualifications and resumes. It's about finding people who are engaged in your business in a fundamental way. The employees who are self-motivated, passionate, and genuinely interested in and excited by your company's goals will be at the heart of its success.

Hire Enthusiasm

The people with the longest resumes are not necessarily going to be the best workers. Most competitive companies need people willing to put in long hours on projects that take real creativity and persistence to bring to fruition. Bill McGowan lived by this lesson. He hired only people whom he sensed were excited by MCI's potential—and, in the beginning, acceptable candidates had to be true visionaries to see beyond the immediate, seemingly insurmountable task at hand. It's been said that MCI management had to be a little crazy to take on the exceedingly powerful Ma Bell. But from the beginning of time, greatness has often been achieved by people whose less-prescient contemporaries judged them to have a loose screw. History books are filled with accounts of explorers, inventors, artists, and builders of every type who bucked popular opinion to achieve success, often at great personal expense.

At MCI bucking popular opinion meant signing on with an upstart company that offered long hours, low pay, and shabby offices. But the drawing card—and one that was eagerly taken up by Bert Roberts, Jerry Taylor, and Larry Harris, to name just a few—was the opportunity to be part of an exciting, innovative, and challenging venture that had a chance to make a lasting mark on the world. As former CFO Bill Conway has said, "I thought it had a

> Newcomers add a different perspective, challenging methods the rest of the organization might take for granted. For example, we might all spend our time shooting alligators, while the newcomer might ask, "why not drain the swamp?"
>
> – BILL MCGOWAN

chance to be one of the great companies in America. I didn't know if it would be, but I said I'd like to be a part of that."

Each of these men—as well as Tim Price, who joined the company later—had a good deal of the entrepreneurial spirit that so defined McGowan. And, as McGowan recognized, brains combined with an appetite for risk was just what MCI needed.

Hire Experience in Key Positions

MCI has always been known as a young company, but its survival at key junctures was directly linked to the knowledge and experience provided by seasoned veterans. Early on, McGowan sought out Tom Leming, Stan Scheinman, John Worthington, and Carl Vorder Bruegge to handle all-important technical, financial, legal, and marketing duties, respectively. All had solid reputations that brought them to McGowan's attention. Leming's experience with microwave technology, Scheinman's knowledge of leasing arrangements, and Vorder Bruegge's marketing skills had been earned at top-notch companies over the years. Later, Orville Wright would come on board to oversee operations and Wayne English would assume the chief financial officer's duties, each of them bringing impressive credentials from their experience at other Fortune 500 companies. Enthusiasm, drive, and imagination combined with a certain amount of seasoned experience will be invaluable to the success of your division or company.

> I thought it had a chance to be one of the great companies in America. I didn't know if it would be, but I said I'd like to be a part of that.
>
> – BILL CONWAY

Action Steps

1. Write out a list of all the reasons (not including tangible pay and benefits) that a potential employee should be excited to work at your company. This list might include such entries as entrepreneurial spirit, encouragement of decision making, creative freedom, cutting-edge technology leadership, and so forth.

2. Ask candidates in interviews to relate at least two new projects, products, services, or programs that they were responsible for creating or implementing, and two they would like to implement at your company.

3. Focus on the candidate's knowledge of current events surrounding your company's business.

Don't Be Afraid to Make Mistakes

With all of its striking success, it's easy to forget that MCI has made its share of blunders—some of which were major missteps. But what sets MCI apart is that it has never let fear of failure keep it from trying new ideas.

Acknowledge and Correct Mistakes Immediately

You won't have to be afraid of mistakes when you know that your liability will be limited by a quick corrective response. MCI's 1972 satellite venture with Lockheed, for example, initially gave MCI much needed credibility with both investors and customers. Within two years, however, as the alliance proved to be a financial drain, MCI knew the partnership wasn't going to work, and it wasted no time in moving to shed its interest.

A more recent example is the decision to pull the plug on 1-800-MUSICNOW. The company had high hopes for this music retailing and on-line shopping service, and why not? Music retailing is a $12 billion business, and home shopping has proven to be a hot property in the 1990s. "You call. You listen. You like. You buy." Or so the company thought, pumping $10 million into the new service. Unfortunately, lots of people called to listen, but few actually bought. So only one year after its November 1995 launch, 1-800-MUSICNOW was no more.

The confidence to risk making a mistake came from Bill McGowan himself. He didn't castigate an employee for trying and failing as long as the person had made a reasonable judgment. "Just as long as you don't make too many [mistakes]," he would say, "if you learn from them, there's no harm." That same attitude prevails among current management. Jerry Taylor believes that too few mistakes belies a too conservative attitude, and Tim Price encourages employees to make risky decisions, saying, "Do

> The day we have made our final mistake is the day we die. Corporations display the same characteristic; they, too, die. But seldom from a mistake; more often from fear of making a mistake.
> – BILL MCGOWAN

it, and know that you're not going to get shot in the back if it goes wrong."

Understand Your Limitations

Many of MCI's missteps could have been fatal had it not understood just how close to the edge it could skate before it needed to change direction. A company willing to take risks must also have the insight to know when to pull back. That insight, coupled with the guts to make the hard choices that sometimes have to be made, is one of the key reasons that MCI managed to survive both its own errors and the adversity that inevitably comes in a free, competitive marketplace.

Stan Scheinman established tough cost-cutting measures in the early 1970s when operating funds were negligible, and the company didn't hesitate to reinstitute such harsh procedures again in the mid-eighties, after increased competition and the burgeoning costs associated with the advent of equal access took their toll on MCI's profits. Cost-cutting sometimes meant layoffs, too, which Bill McGowan hated. But he never lost sight of the fact that unpalatable choices could not be avoided when the company's very survival was at stake.

> The simple truth is that early on we didn't know what we were doing half the time. And the other half of the time we were dead wrong.
>
> – BILL MCGOWAN

Action Steps

1. List three mistakes or miscalculations made by your division or company in the last five years and how quickly these missteps were abandoned or rectified.

2. Scan your current projects right now. Search the horizon for the losing propositions—high launch costs versus low earnings expectations, time-eater agendas with no foreseeable benefits, ill-conceived designs for soon-to-be-obsolete products.

3. Discuss with your colleagues the best-case and worst-case scenarios for these projects and clock how fast you can get them fixed or killed.

Living Lesson 3: Empower Your Employees

Make employees feel their worth to the survival and success of their company by delegating decision-making power to them and making them accountable for profit and loss. This policy will effectively produce a motivated workforce and top performance.

Delegate or Die

MCI's strong belief in delegating responsibility can be traced to Bill McGowan. He had so much to accomplish in his battle with AT&T that it was either delegate or die. But even if delegation hadn't been crucial to MCI's existence, McGowan was the kind of manager who firmly believed that you hire good people, then get out of their way.

MCI still adheres to that concept today. Top management, for example, doesn't even see the company's advertising spots before they go on television. Of course, management is very much involved with determining what the message will be, but how the advertisements say it is left entirely to the staff.

Jerry Taylor believes so strongly in delegation that he has been known to take off on vacation right before a key launch.

> For the amount of things we get done, and the output we have, and the growth we have, you cannot possibly afford to have managers who want to do everything themselves.
>
> – ANGELA DUNLAP

Action Steps

1. Delegate to others three projects you were going to do yourself.

2. Ask employees for three new ideas on how the company can become more profitable or more efficient. Choose one idea from each person and make that person responsible for its implementation.

3. Assess the effectiveness of and respond to employees given responsibility. Reward those who perform well and reevaluate the shirkers.

Prior to the introduction of Friends & Family, Taylor surprised his staff by leaving for to the Bahamas. He couldn't even be reached by telephone. Taylor's action, however, was not irresponsible behavior. Rather, it was a way to say to his staff, "I trust you. You can do it." It goes without saying that Taylor would never jump ship without making sure that other senior managers were available in case of serious problems, but by giving people responsibility, he encourages accountability.

Living Lesson 4

Communicate With Your Employees

To keep employees connected to the enthusiasm they have for the company, you must stay in touch with their needs and concerns and establish yourself as an immediate presence in their lives. MCI fosters this relationship through constant communication. Workers must know that management hears them and is responsive, and that their decisions count. With a work force of 55,000 people, this task is not an easy one. But management is constantly working on new ways to touch base with employees—through video presentations, informational seminars, and frequent visits to different MCI locations.

These visits are what employees respond to most. It gives them a chance to ask questions and get face-to-face answers. Typical of MCI's freewheeling style, these meetings are casual and even entertaining. (Bert Roberts, an amateur magician, likes to entertain the troops with magic tricks.) This may seem a simple concept, but when employees see that the top brass have a sense of humor and interests outside the company, they get a real sense that they are working with someone, not for someone.

> Empower and encourage young employees so that they aren't afraid to make decisions. I worry less about employees who make the wrong decision than about those who are afraid to make a decision in the first place.
>
> – JERRY TAYLOR

Action Steps

1. Hold a state of the union address for your division or company at which you reaffirm the goals that have been set and relate the wins and losses of the last six months. Be sure to share a part of your outside-the-office self with the group.

2. Open the floor to questions. If the group is shy to begin with, have some questions prepared that you know address some of the needs and concerns of all involved.

3. Walk around the office once a month and stop off at people's cubicles or offices to ask how things are going.

Push the Envelope

The amount of time and energy needed to complete any given task is always dependent on who is doing the task—some will need more time, some less. It is the leader's responsibility to set the pace and push people to excel beyond their normal limitations.

Set Near Impossible Goals

> Remember—it's not bragging if you can do it.
> – BILL MCGOWAN

When an important project needs to be done quickly to gain an advantage over the competition, try setting a near impossible deadline to really charge up the troops. MCI has had to do the impossible on many occasions in its fight to gain a foothold in the closed telecommunications industry of the 1960s and 1970s. Perhaps the most remarkable example of this was the construction of its first microwave network in 1973. Under a conventional schedule, this task would have been impossible. But the construction crews forged ahead anyway, forgetting to worry about the conventional wisdom.

Individual reactions to such impossible goals will show you who has the strength of character to take the plunge against big odds and who simply wants an easy ride. Achieving such goals will also energize those involved in the effort and carry over to raise morale throughout the division or company.

> The point is that nothing is impossible. Some things just haven't been done yet.
> – BILL MCGOWAN

Encourage Friendly Competition

To move at the rapid pace that is required to stay competitive, MCI advocates a Darwinian management style: Three or four employees (or departments) are given the same task, and whoever gets the job done first, best, and cheapest wins.

McGowan was a big believer in this method, but not because he wanted to pit employees against one another. He never advocated internal discord. However, knowing that MCI had to fight hard and move fast, he reasoned that creating informal intra-office competition was the best way to get things accomplished quickly and competently.

Consider the development of the company's fax service. McGowan gave the assignment of developing a fax product to no less than five different groups. Among them was Tim Price, who describes his relative position to McGowan at the time as distant: "I not only didn't report to him, I was about four levels down."

McGowan, having heard that Price was a mover, picked him along with the four other employees to plan the fax rollout. McGowan's approach request was simple: If you can't do it in forty-eight hours, let somebody else tackle the project. Price didn't roll out MCI's fax service in forty-eight hours, but it was operational in record time nonetheless. The project was conceived, designed, and launched within a month.

> The whole history of MCI has been a history of meeting challenges, no matter how improbable the odds.
> – BILL MCGOWAN

Action Steps

1. Take your second most important project and give it to three different individuals or teams to complete, making certain they know they are in a competition.

2. Set an ambitious deadline to see the responses of those faced with such a challenge.

3. Reward the individual or team that perseveres against the odds and completes the project in an acceptable manner first.

Living Lesson 6: Stay Flexible

The Information Age demands flexibility. It constantly changes the terrain that a successful business must traverse with innovations in technology and revolutionary new methods of performing old tasks. The smart manager will stay pliable and resiliant in the face of these innovations, ready to move at a moment's notice today into promising new opportunities that may not have even existed yesterday.

Travel Light

Bill McGowan recognized that to stay competitive in a world marked by rapid technological change, MCI had to be able to shift smoothly and quickly from the outmoded to the up-and-coming. That's why the company never invested in manufacturing plants; it was one of McGowan's key strategies for beating AT&T to market.

He knew that AT&T was saddled with huge investments in its Western Electric manufacturing subsidiary, which made it difficult to switch to newer, more profitable technologies. MCI, on the other hand, could move from technology to technology—microwave, fiber optics, satellite, and back to fiber optics again—abandoning those not suitable and embracing those it wanted with a minimum of cost, time, and corporate angst.

Let the Customer Drive Innovation

MCI has also been able to stay flexible as well as attentive to customers—because technology does not drive its product development—the customer does. AT&T developed technology through its Bell Labs division, then brought it to market. Conversely, MCI listened to what consumers wanted, then went to the technology community to see who could provide the best product for

> Whenever I had to make a decision, it was very simple. I figured out the way AT&T did it and I did exactly the opposite and it always worked.
>
> – BILL MCGOWAN

the least amount of money and in the fastest time. As a consequence, MCI became known for its technological innovations, and today has the world's most modern network.

Be Prepared to React Quickly

Flexibility also results from a management style that promotes fast decision making, rather than long, drawn-out processes that drain precious reaction time. A prime example of MCI's agility was Roberts's decision to go with Friends & Family. After the three-hour meeting with Taylor and Price, Roberts asked a few questions about the product and quickly approved it—with the proviso that it be rolled out in one month. There were no reviews, no committee meetings. Roberts knew that his best and brightest were working on the project.

The ability to act quickly and decisively has been in evidence in the many alliances MCI has moved into and out of throughout its history. It had the adaptability to partner with corporate giant IBM, for example, utilize the entrée to large accounts that Big Blue afforded it, and then, three years later, shake hands and move on.

> We're as panicked as we were fifteen years ago. Maybe the things that bother us are different, but you need to be panicked to survive.
>
> – BERT ROBERTS

Action Steps

1. Identify the largest fixed-cost assets of your division or company such as highly capitalized manufacturing lines. Are there outside companies that could perform the function of these lines more inexpensively? Could switching give you more flexibility in upgrading to new processes or technologies?

2. Call every client or customer in your database and ask them if there is any service or product they wished were available. Do you have the flexibility to meet their needs?

3. Put the most promising ideas into immediate development and track the time it takes from conception to launch. Try to improve turnaround time on the next product developed.

Diversify Your Workforce

Living Lesson 7

Beware of the tendency to hire people who look, talk, and act the same way you do. The cookie-cutter approach can lead to a stagnant organization. Diversity encourages new ideas and ways of doing things that can lead to new breakthroughs.

Hire Diversity

> Divestiture is the most dramatic watershed in this industry since Theodore Vail decided that being regulated would be nice work if he could get it.
> – BILL MCGOWAN

No two men could have been less alike than McGowan and Jack Goeken or, for that matter, McGowan and Orville Wright. Yet McGowan brought Wright on board as the company's president and chief operating officer because he knew Wright's staid, detail-oriented approach would complement his own flamboyant style. Such creative *incompatibility*, in fact, played an important role in MCI's growth.

What Taylor describes as "automorphic hiring" has never occurred at MCI because the company works hard to guard against it. "Cultural diversity is a very, very real strength," he says. Recognizing the importance of diversity is one reason that MCI places such a high value on change—constantly moving employees and restructuring the organization.

Action Steps

1. Survey the demographics of the people in your division or company. What are the relative percentage breakdowns of age, gender, ethnicity, sexual orientation, race, religion, and socio-economic class?

2. Target the lowest represented groups for favorable weighting when the next round of job applicants are interviewed.

3. Instruct human resources to search actively for the best candidates in these groups so others will have no reason not to hire diversity.

This policy of inclusion and recognition of differences means that all employees feel comfortable with their role at MCI, and no one feels intimidated by the people at the top. They can go about their responsibilities with self-confidence in the knowledge that they will be judged on their merits.

A culture of diversity and change also fosters corporate camaraderie. Departments never get the chance to become isolated or develop an intracompany us-against-them mentality because the them and the us is constantly changing. At MCI employees are comfortable offering suggestions to other divisions in the organization. In fact, all employees are free to send e-mail directly to Bert Roberts, who gets messages about marketing ideas from people all over the company.

MCI's insistence on diversity doesn't stop with people who have different ideas. Thirty years ago, before the terms diversity and pluralism became common in business parlance, McGowan made a habit of hiring people of varying ethnicities and backgrounds. Besides being good for internal morale, a diversified workforce helps MCI stay in touch with its customers, who, of course, are not all middle-aged white males.

Form Alliances

From its inception, MCI has used corporate alliances to get from point A to point B in the shortest possible time. Any alliance must be considered carefully from all sides, of course, but don't let your possessiveness or fears overshadow the benefits that can come from a good partnership. MCI's record highlights three excellent ways that a big ally can help: in increased prestige, in expanding market reach, and in raising capital.

> I've noticed over the years that if you ask a monopoly if the glass is half empty or half full, they demand the whole glass.
>
> – BILL MCGOWAN

Increase Prestige

MCI has gained instant credibility in two notable alliances: With Lockheed in 1971 and later with IBM in 1985. Each alliance came at a critical time in MCI's growth.

In the case of the Lockheed joint venture, the plan to build a satellite system never actually proved to be profitable, but it did catapult the company to a higher corporate level. The Lockheed association effectively removed MCI's image as an upstart. Indeed, shortly after MCI signed the agreement, it was able to place a series of private stock offerings that brought in about $5 million.

This strategy would again prove successful much later in its history when MCI relinquished 16 percent ownership in exchange for IBM's Satellite Business Systems (SBS) division. Although some MCI executives feared IBM was positioning itself for a takeover, outsiders—and McGowan himself—saw it as a magnificent coup for MCI. MCI needed the respectability that IBM provided. Despite the company's success in the financing arena, it still lacked marketing credibility with some of the larger accounts. At the time, large corporations such as Johnson & Johnson felt uncomfortable about switching to MCI, which was still fifteen times smaller than AT&T.

How did MCI manage to attract IBM? MCI's management realized that it could offer something IBM wanted: an exit from the telecommunications business. "IBM believed that telecommunications services could be purchased inexpensively," Roberts explained. "Its interest was in the large mainframe computers and revenues those computers provided."

Besides, SBS had lost nearly $400 million in the three years before its sale to MCI. But this fact didn't disturb MCI because management felt so strongly about IBM's ability to increase the company's visibility in the business and investment communities. (One month after the alliance, MCI raised $575 million in bonds—something it had not been able to do for two and a half years because of its compromised position.) Moreover, SBS did come with some very specific benefits. It brought with it an additional 200,000 long-distance customers; new capabilities in video-conferencing, high-speed faxes, and computer information services for business customers; and data and voice networks for large corporations.

Although the alliance would last only two years (IBM sold back its MCI stock in 1988), MCI considered the experience an unequivocal success. The temporary partnership had given MCI exactly what it needed to move ahead.

> The RBOCs are so big, calling them "Baby Bells" is kind of an oxymoron, sort of like jumbo shrimp.
>
> – BILL MCGOWAN

Expand Market Reach

Reaching the global market was the immediate benefit to MCI from the 1993 strategic partnership with British Telecommunications (BT), the fourth-largest carrier of international traffic. At the time of the agreement, MCI had nearly $12 billion in revenues and was the sixth-largest carrier. The press characterized the marriage as the first telecommunications "mega-alliance," and enabled it to offer global service before AT&T.

In planning its strategy for obtaining an international partner, MCI considered many factors, not the least of which was finding a company that could provide an entrée into the Asian market. But Roberts cited the many commonalities between MCI and BT—including language—as a factor that helped management arrive at a decision. Of considerable importance was BT's success in the United Kingdom, a notably competitive marketplace.

Management Strategies

Raise Capital

The other more obvious benefit to MCI from the BT merger was the $4.3 billion cash infusion it received in the deal, which gave BT a 20 percent interest in MCI. Thus, the BT arrangement not only extended MCI's international reach without requiring it to make a huge capital outlay, but also provided MCI with cash to expand its presence in North America, still the fastest-growing telecommunications market in the world.

Of course, many analysts voiced the same concerns as they had after the 1983 billion-dollar offering. The company, they claimed, was overcapitalizing—not to mention diluting ownership by bringing in such a big investor.

But MCI recognized that in an industry as capital-intensive as telecommunications, it was virtually impossible for a company to have too much money. Particularly in this era of increasingly rapid change, having cash available gave management the ability to ride the waves of change and benefit from the opportunities provided by new technologies and methods, rather than watching from the sidelines as it hunted for cash.

"That capital gives us enormous leverage capability," Roberts says. "What a mistake it would be for me as chairman and CEO if somebody ten years from now says that MCI missed an opportunity because it didn't have the financial resources."

> MCI really was a creation of AT&T. We wouldn't have had a prayer had AT&T played by the rules.
>
> – BILL MCGOWAN

Action Steps

1. Cultivate relationships with companies that complement your products or services or that are in the same field but different markets.

2. Periodically assess the benefits and drawbacks of strategic alliances with other companies.

3. Focus on struggling divisions of other companies or competitors, which you may be able to acquire cheaply. Would they find a more profitable fit at your company?

Map Out the Regulatory Landscape

Living Lesson 9

Jack Goeken once reflected that had he fully anticipated the extent of the wranglings with the FCC, he might never have entered the fight against AT&T. Certainly, no one can ever accurately predict the future, but Bill McGowan made sure going in that he knew all there was to know about the telecommunications industry and the workings of the Federal Communications Commission.

Do Your Homework

Whether it's the maze of rules of your local city hall or the massive red tape of a full-blown federal bureaucracy, you must do your homework to understand the rules and regulations involved with your products or services. Bill McGowan, for example, studied everything he could get his hands on before he even agreed to join MCI. He also talked with AT&T executives, FCC regulators, and officials at industry associations. This kind of thoroughness not only set McGowan apart from his competitors, it gave him the tools he needed to win the federal and state regulatory rulings that allowed MCI to grow in the face of AT&T's unrelenting opposition. He used his extensive knowledge to carefully plan his appeals to the FCC so as to assure that MCI could get its foot in the door of the long-distance telephone market.

> June in Washington [D.C.] is beautiful. The cherry blossoms have all gone, but Congress is there to provide around-the-clock entertainment.
> – BILL MCGOWAN

Take Advantage of the Law

It's always a smart move to tailor the company's structure and activities to take fullest advantage of the law. For example, when McGowan first came on board, he pushed to form an umbrella company made up of several mini MCIs that would be linked together to carry nationwide telephone traffic. Among the reasons for this setup was his realization that the FCC always leaned toward granting radio and television licenses to local companies. The original Chicago-to-St. Louis route was excluded from the new entity because management knew that such a move would require FCC action, and that, in turn, would mean delay.

Furthermore, McGowan planned to structure the network as a group of independent franchised companies, which meant that he, as the franchiser, could handle multiple FCC filings, reducing the time and expense of individual applications. McGowan worked to turn regulatory quirks to the company's advantage.

> Any time you have 40 percent of the Congress supporting a bill to put you out of business...you have one helluva problem.
> – BILL MCGOWAN

Scrutinize Public Documents to Stay Informed

Despite AT&T's long-standing relationship with the FCC, one in which the commission had effectively become AT&T's protector

Action Steps

1. If you don't know what federal, state, and local regulatory agencies your company reports to, find out from the person who does. Some likely candidates are the owner, the CFO, the COO, the chief administrator, or the office manager. In all likelihood, your competitors will also be reporting to the same agencies.

2. Request information on your competitors from regulatory agencies such as the Securities and Exchange Commission (SEC), the Internal Revenue Service (IRS), U.S. Patent Office, the Food and Drug Administration, the Board of Governors of the Federal Reserve System, Federal Energy Regulatory Commission, Environmental Protection Agency, the Federal Communications Commission (FCC), U.S. Interstate Commerce Commission, the industry analysts at the U.S. Department of Commerce, the state Office of Corporations, the state Office of Uniform Commercial Code, the state Securities Office, state Department of Commerce, the county government, local courts, local chamber of commerce, and local development authorities.

3. Expand your information search into nonregulatory sources such as trade magazines, the business or managing editor at your local newspaper (companies love to brag about future plans), and your competitors' suppliers.

in an effort to foster the best possible telephone service for the American consumer, the long-distance giant still had to file its plans with the FCC just like everyone else. So, to take advantage of what he saw as yet another potential plus for MCI, McGowan set up MCI's offices in Washington, D.C., just down the street from the FCC. This made it easy for MCI's attorneys to walk to the FCC to scrutinize AT&T's long-distance rate filings, gaining insight into its pricing.

Shape the Regulatory Landscape

Bill McGowan's awesome skill at playing the regulatory game is so renowned that, according to *The Washington Post*, lobbyists describe winning a long-shot regulatory victory as "pulling a McGowan." He knew better than most that companies live and die by government decisions. Witness the effect on the pharmaceuticals industry when President Clinton began his drive to reform health care. With drug companies facing a potential squeeze on profits, it didn't take long for stock valuations to drop by some $70 billion—and no legislation had even been passed.

Even a suggestion of a major regulatory shift can have a dramatic effect on the stock market—to which MCI can readily attest. Three weeks after its historic billion-dollar offering, its stock plummeted 26 percent in one day when the FCC announced an interim decision that would raise access charges to specialized common carriers.

> Once people know and realize that all that's standing between them and convenience is government regulation...either the regulation will change or the government will.
>
> – BILL MCGOWAN

When Necessary, Litigate Your Cause

By placing himself squarely in the center of Washington, McGowan was able to exert as much influence as possible. He used a variety of tactics over the years, but his course of action in the early days was litigation. In fact, if it had not been for McGowan's successful use of litigation, the telecommunications industry would be very different today. To chronicle each instance in which MCI used the law to further its cause would take volumes, but it's important to understand MCI's three basic tactics in these cases.

1. Hire experienced spokespeople. MCI hired lawyers who had experience obtaining approvals from state and regulatory bodies. When it could, it hired ex-FCC lawyers, reasoning that if anyone could understand the byzantine workings of the commission, it would be a former employee. Among McGowan's early hires was Kenneth Cox, whom he plucked directly from the FCC in 1970. It was Cox who helped orchestrate MCI's strategy leading up to the AT&T antitrust proceedings.

2. Be creative in your campaigns. MCI often had to be imaginative and daring in its tactics. Once when AT&T dragged its feet on providing the FCC-ordered interconnections to MCI, McGowan wanted to sue, but he knew that he didn't have the money to both build a network and pay for expensive litigation against his deep-pocketed rival. That's when Cox suggested that MCI use a backdoor approach, which included sending letters to the Justice Department's Antitrust Division detailing AT&T's tactics.

Although MCI eventually filed a civil antitrust suit because the government was moving too slowly, the letter-writing campaign resulted in a congressional hearing at which McGowan passionately presented his case and suggested the breakup of Ma Bell. Cox's inventive approach worked to create sympathy for the MCI cause, not only on Capitol Hill but with the general public as well.

The strategy of engaging both public and private support has always played a crucial role in MCI's ability to influence legislation. McGowan figured that if you applied enough pressure, something would eventually give.

To prove to Washington that MCI's pleas were valid and not merely sour grapes from a tiny competitor, McGowan thought it essential to gather grassroots support. To that end, he sent out twenty-five thousand letters to friends, customers, and associates, pleading his case and urging them to help topple AT&T's monopoly.

This kind of old-fashioned maneuvering to provoke a groundswell of support is something MCI still uses. In its recent battle to win the last available satellite license from the FCC, Larry Harris employed the tactic with great success. Cable television magnate John Malone had an option on the license. When the license expired and the FCC chose not to renew it, Harris saw his opening. He began sending the agency a series of letters

> Being rescued by [politicians in] Washington is like being lost at sea and getting picked up by the Titanic.
>
> – BILL MCGOWAN

Management Strategies

indicating, not so subtly, that if only the satellite were up for auction, MCI might be willing to make a substantial bid. The agency was persuaded and decided to auction the satellite license. After nineteen rounds, MCI won with a bid of $683 million.

Harris also helped to coordinate a massive letter-writing campaign from MCI employees and their families and friends to win as many protectionary measures as possible in the Telecommunications Act.

> We don't award pizza monopolies just because it might be neater to have only one set of carry-out trucks on the road.
>
> – BILL MCGOWAN

3. Make friends and influence people. MCI always recognized the importance of the right Washington contacts. Besides gathering broad-based support for its causes, MCI also carefully cultivated the legislators themselves. McGowan had a close relationship with Speaker of the House Thomas P. ("Tip") O'Neill, the longtime Massachusetts congressman, and he worked to hire people who were already wired into Washington.

One of these insiders was Eugene Eidenberg, who succeeded Cox as head of regulatory affairs. Before joining MCI, Eidenberg was chairman of the Democratic National Committee and a fellow at the Institute of Politics at Harvard's Kennedy School of Government. He also served in President Jimmy Carter's cabinet. Eidenberg clearly knew his way around Washington and was keenly aware that schmoozing was the modus operandi of Capi-

Action Steps

1. Analyze the intelligence you've gathered on your competitors from the regulatory agencies and other information conduits you've contacted.

2. Identify specifically the advantages in patents, licenses, permits, zoning, product specifications, outlet locations, and so forth, and choose the most important one as your target.

3. Do whatever it takes to change the law to your advantage on this target issue by networking with state, local, or federal representatives; mounting a publicity campaign to use public opinion to effect change; or litigate your cause in the appropriate court.

tol Hill. He spent countless hours wining and dining officials who might provide support.

MCI's success in influencing legislation is grounded in its belief that there is a place for both litigation and cultivation. Companies without the legal and financial resources to instigate lawsuits may still be able to exert power in other ways. No matter which avenue you pursue, the key is realizing that government ultimately has the power to determine a company's fate.

Chapter 10

Marketing and Advertising Strategies

You can't run a successful ice cream business today if you only offer a vanilla flavor.

— BILL MCGOWAN

Man: Have you been talking to our son long distance again? **Woman:** (Nods and whimpers)

Woman: Have you seen our long distance bill?

MCI
The nation's long distance phone company
1-800-624-2222

Announcer: Sure, reach out and touch someone. Just do it for up to 30, 40, even 50 percent less.

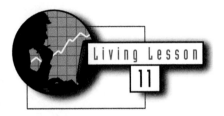

Create Brand-Name Recognition

Living Lesson 11

You may think that your product or service is too small, too niche oriented, or too plain to create brand-name recognition. But branding a commodity simply means giving it a personality or image that generates a feeling of trust in the consumer. Prior to the advent of Friends & Family, for example, long-distance telephone calling was merely a price-oriented commodity product. People could recognize the company names—AT&T, MCI, Sprint—but there were no brand names for the products of these companies. Competitive advantage was based primarily on which company offered the cheapest calls. MCI changed all that and in the process revolutionized the concept of advertising.

> Even if you are one of thousands in a large company, you can achieve greater success by eliciting brand recognition of the work you or your team generate in the form of a reputation for excellence. By consistently delivering outstanding performance, you can raise your name to a kind of brand recognition where people associate it with reliable, quality service.

Advertise Your Product or Service

To create brand-name recognition you must get the word out about your product or service in a way that makes it stick in people's minds. Bob Schmetterer, a partner at Messner Vetere Berger McNamee Schmetterer, MCI's advertising agency, explains that Friends & Family is a "brand in the Procter & Gamble sense of the word. It means something unto itself; it has a soul, a persona." People feel something about the product and differentiate it from other offerings; the audience has something to grab on to. And that is what makes the Friends & Family marketing strategy unique.

You don't need to spend millions of dollars on television advertising, however, to achieve brand recognition. A catchy yellow pages or local newspaper advertisement is a low-cost way of starting a brand image. Another great advertising method is to get your customers talking about the quality of your product. Word of mouth is essential to solidifying the brand in people's minds. But perhaps the best low-cost path to branding your product is through publicity. Sponsoring charity events or community awards, holding tent sales or promoting on local talk shows are inexpensive ways to garner high visibility for your product.

Schmetterer points out that Procter & Gamble built its history and its business on its ability to give its products personality. Lots of companies produce soap, but few have achieved the name recognition and staying power of Tide, Cheer, and other Procter & Gamble brands. In that same tradition, MCI has proven itself as an excellent marketer and developer of brands.

Other companies have since tried to brand their telecommunications products, but none has been as successful as MCI. No competitor's product consistently generates the kind of response from consumers that Friends & Family does. Opinions about the product may be positive or negative, but virtually everyone knows what Friends & Family is, Schmetterer points out.

Today, with cable companies, satellite companies, broadcasters, the regional Bell operating companies, and the traditional long-distance carriers all competing for the same territory, the ability to delineate a product assumes even greater importance in the telecommunications industry.

> Branding sets MCI apart. it is the only company I know...that has actually developed meaningful consumer-based brands....
>
> –BOB SCHMETTERER

Action Steps

1. Choose a target name to mount your campaign behind—the name of your company, a product, a line of products, a service, etc.

2. Determine the available avenues of getting the word out—advertising, telephone calling, local talk shows, sponsoring events, company newsletter, word of mouth. Explore the ways in which your commodity-oriented products could be branded, from surface changes like a new name to more substantial ones, like adding a new valuable perk.

3. Coordinate the various elements of your campaign into a blitz that takes place over a specific period of time, where each element reinforces the others. After the blitz period ends, solidify your gains with continuous advertising and publicity.

Make the Billing Process Easy

Billing is one of the four cornerstones of MCI's highly successful marketing strategy. It is the leader in billing practices for the telecommunications industry and the only company to provide customers with a single bill for all of its services. This type of billing instrument is difficult to maintain, which is attested to by the fact that AT&T, Sprint, and the others still haven't been able to pull together a similar method of billing for their services.

Build Dependable and Accurate Systems

Billing is a perennial problem that all of the telephone carriers, including MCI, have experienced. Nothing makes a customer angrier than an incorrect bill. Consumers often react by immediately switching service to a competitor. So, poor billing practices are more than an annoyance to your customers; they can be very costly to your company.

Make It Simple

Get all your products and services on one succinct, clearly understandable statement. Nothing delays payment like not knowing what is being paid for.

Prepare for Success

Anticipate sales volume of new products to be offered with updated and more powerful billing systems. MCI learned its lesson about billing the hard way during two of its most important successes. During the growth years of Execunet, as sales increased, MCI's internal support organization couldn't keep pace. During the 1970s, employees prepared bills manually with the occasional assistance of ancient IBM tabulators. Typists corrected their mistakes with erasers and correction fluid. The product of their efforts often were bills that were unreadable.

The entire billing process was a mess from start to finish, with costly delays that meant reduced cash flow for the company, and—most dangerous—irritated customers. MCI improved the situation slightly by hiring temporary employees, but error rates remained high. Finally, MCI began the long and arduous task of sorting out its billing problems. Over an eight-year span from about 1978 to 1986, it invested in systems to address its billing problems which led to great improvement.

But in 1986, when a residential sales campaign flooded the company with new orders, MCI saw its billing problems begin to worsen again: After the residential sales campaign flooded the system, MCI estimated that it was losing $100 million annually because of billing glitches.

In the end, MCI improved its system by installing a more up-to-date information system at a cost of $300 million. Finally, by 1988, on-time billing accuracy reached 100 percent, and it has stayed close to that level ever since.

> After the residential sales campaign flooded the system again, MCI estimated that it was losing $100 million annually because of billing glitches.

Action Steps

1. Get a few sample copies of the bills sent out to your customers. Ask different groups of people, including a few typical high school kids, to look them over and circle the products purchased, the itemized amounts, and the total to see if the bill format is simple enough.

2. Determine the average turnaround time from product delivery to billing. Is it fast enough? Can it be improved?

3. Determine the average turnaround time from billing to payment receipt, and compare to a reasonable industry standard. Are you being paid as quickly as your competitors? Are there any ways to accommodate the customer such as offering to bill in different languages (Spanish, for instance)?

Tailor Your Marketing

Living Lesson 13

To be most successful in business, whether your company serves a local, regional, national, or global market, you must tailor your marketing to embrace the diversity of society. To accomplish this, you must begin by determining the most productive market segments for your company's product or service and targeting them with specialized campaigns. MCI, for instance, has always recognized the value of the diverse segments of American culture. Accordingly, it markets to them directly and aggressively.

Target Specific Geographic Areas

The equal access campaign is a good example. For each specific segment of the long-distance market, MCI constructed a different approach to gain customer share. When the company found that a higher proportion of customers in Charleston, West Virginia, and Minneapolis, Minnesota, were able to choose their long-distance carrier, it drafted tactics aimed specifically at each of those two markets.

> On calls to Taiwan, MCI carries about 50 percent of all calls originating from the United States because we were the first to approach the Chinese-speaking market in its own language....
>
> JOHN DONOGHUE

Say It in Different Languages

Another example of successful market targeting is found in MCI's focus on diverse ethnic groups in the U.S. It markets the Friends & Family product in nineteen different languages within the United States.

"We saw in the early nineties that international minutes were growing about three or four times faster than domestic minutes," related John Donoghue, senior vice president of consumer marketing. Furthermore, the international minutes had three to four times greater margin. And not surprisingly the largest concentration of international callers was among recent immigrants for whom English was not their first language.

With that information in hand, MCI started aiming its Friends & Family message at specific ethnic groups. "We were able to pick up about 50 percent market share in some countries," Donoghue said. "For example, on calls to Taiwan, MCI carries about 50 percent of all calls originating from the United States because we were the first to approach the Chinese-speaking market in its own language, using imagery and culture to talk about what that market cared about—international calling."

> You can't run a successful ice cream business today if you only offer a vanilla flavor.
>
> – BILL MCGOWAN

Research Sales in Diverse Categories

Tailored marketing may target such diverse categories as geography, ethnicity, race, religion, sexual preference, gender, age, and socio-economic status, among others. To know where you're weakest and where the greatest potential for growth is, you must research, either before product launch or after the latest sales figures are in, how each diverse group feels about your products or services.

Action Steps

1. Break down recent sales figures by as many categories as possible, including those mentioned above.

2. Determine the areas with the strongest potential for growth. Are they addressed in your marketing plan? Does their potential justify more marketing attention?

3. If uncertain, try experimenting with low-cost advertisements in a few niche publications and gauge the response.

Living Lesson 14

Leverage Off the Size of Your Competitor

Corporate jujitsu is an effective way to use an adversary's size to your own advantage. "The best way to do that is through speed of execution," said Tim Price. "The smaller company can react and bring things to market before the larger player can."

Beat Your Rivals to the Market

MCI has effectively employed this tactic many times against its huge rival AT&T. With its smaller size and its aggressive, entrepreneurial style, MCI has been able to introduce major new services to the market with remarkable speed. Take 1-800-COLLECT. In just seventy days, MCI initiated and launched this discount collect-call service that allows customers to access MCI directly, by-passing other long-distance carriers and their higher usage charges. By the time AT&T jumped in with its copycat 1-800-OPERATOR product, MCI had already staked out the dominant share of this niche market.

MCI also brought to market networkMCI, the company's innovative suite of integrated communications services for business users, in just 110 days, an unheard-of time considering the complexity of the high-tech offerings.

> Two years from now I hope to reach a level of business equal to Bell's annual bad debt provision.
>
> – BILL MCGOWAN

Leverage in the Numbers

In addition to speed, the greatest advantage a smaller company has over a larger one is in the benefits of discounting. When AT&T would consider matching an MCI long-distance discount, for example, it knew that it would benefit much less in terms of increased market share and visibility because it already had the lion's share of the market. While MCI signed on more and more customers, the giant AT&T could at best try to tread water.

Marketing and Advertising Strategies

Create a Product That Large Competitors Cannot Compete With

MCI executed this tactic brilliantly with its Friends & Family product. Because AT&T already had 80 percent of overall market share, most AT&T customers were calling other AT&T customers. There would be no reason, therefore, for AT&T to offer a discount on established calling circles. MCI used this circumstance to offer a large discount to those who could establish a calling circle of family and friends. Any loss in per-call revenue was more than made up for in increased market share, free marketing by customers themselves, and the furthering of MCI's image as a quality discount leader.

> When AT&T...is quoted...as saying that MCI can't compete with AT&T, we put that quote on every bulletin board... in the country. It's an incredible motivator.
>
> – TIM PRICE

Stir Up Employee Enthusiasm

There's yet one more dividend for the small-guy competitor—underdog motivation. When Goliath AT&T attacked, MCI employees reacted in true David fashion as if they had been assaulted by a bully. These attacks would bring the troops solidly together against the nasty predator. "What helps us is that we have an enemy to focus on," Price said. And not just any enemy, but one so big and formidable that the stronger the attack, the more enthusiastic the response from the MCI troops.

Action Steps

1. Determine one product category in which a competing company has dominant market share.

2. Calculate the break-even point on your product at a discounted rate and add in a slim profit margin, factoring in the benefits of increased units sold (and decreased per-unit production cost due to larger volume if applicable).

3. List marketing tactics that your competitor would find difficult to copy and choose one to implement immediately.

"When Joe Nacchio (president of AT&T's consumer division) is quoted in *USA Today* as saying that MCI can't compete with AT&T," Price went onto say, "we put that quote on every bulletin board in every one of our customer service centers and sales centers in the country. It's an incredible motivator."

Break the Mold With Your Advertising

In 1980, when MCI was beginning its pursuit of the long-distance residential market, it needed to call attention to itself. Not only was it a relative unknown in the marketplace, its prime competitor was—and still is—one of the most advertised companies in the world. (AT&T runs neck and neck with McDonald's.) With the help of New York's Ally & Gargano ad agency, MCI decided to hit AT&T where it was vulnerable—price.

Throw Out the Rules

The famous split-screen running-meter spot ("You haven't been talking too much, you've just been paying too much") marked MCI's first experiment with this hugely successful marketing approach. The accompanying print ads which spoofed AT&T's own slogan ("Reach out and touch someone, but for a lot less money") were the perfect underscore.

By today's standards, those ads may seem less than revolutionary, but when they first appeared, the advertising game was still a genteel sport, in which even naming your competition was considered out of bounds. MCI needed to get on the radar screen, however, and the way to do that was to make a bold entrance, which it did with a flourish. Those first ads were so successful in the Denver market, where MCI first tested residential service, that it canceled a planned three-month trial period to launch the residential concept nationally.

This brazen approach would develop into MCI's unique marketing style, and one of the most successful campaign records in advertising history. Who could forget the "momma's tears" TV commercial? The genius of the ad was its combination of direct attack and humor.

Establish a Personality

The other ingenious aspect of MCI's advertising has been that it has directly reflected MCI's corporate personality, which in turn has become recognizable to the public. This was exactly the aim of the campaign which featured comedienne Joan Rivers. Rivers' in-your-face attitude perfectly mirrored MCI's own style.

In her spot Rivers gave the audience her characteristic admonition to "grow up," adding that AT&T's operators don't say "Hello, thank you for calling, and have a nice day" because they love you. They do it, she said, "because they're afraid you're going to call MCI." Once again, MCI's advertising reinforced its own image as brash and AT&T's as benevolent and slow.

Be the Nice Guy

Of course, advertising doesn't have to be abrasive to break the mold. In fact, bold advertising can backfire if you're trying to attract an audience that doesn't respond well to brashness. While retaining its image as a tough competitor, MCI has altered its advertising over the years to appeal to different markets. Its Friends & Family campaign is the best and most obvious example. The company gave itself a personality makeover by positioning MCI as a friend that helps you stay in touch with loved ones.

Recruit Your Ad Agency to Your Vision

No discussion of MCI's advertising would be complete without recognizing the unusual relationship it has with its advertising agency. Tom Messner and his partners at Messner Vetere Berger McNamee Schmetterer Advertising Agency, the New York-based firm that handles MCI's business, have played pivotal roles within the company.

A copywriter at Ally & Gargano when that agency had MCI's account, Messner has been around since the early days when Jerry Taylor and Bill McGowan were essentially writing ad copy themselves. He has been personally responsible for some of the

> [McGowan] loved talking about the [commercials] and showing them. And by the way, he loved awards, which is very funny to me, because the cliché in the ad business is that creative people—the art director, and the writer—salivate over these things that clients don't care about. Bill just loved awards.
>
> – TOM MESSNER

company's most famous spots. In many ways, MCI and Messner's agency have grown up together.

Taylor unhesitatingly calls Messner a genius and speaks as if he and his partners were trusted company employees rather than outside vendors. They are included in strategy sessions where ideas are freely exchanged and the ad team's opinions are respected.

Find an Agency That Really Knows Your Company

Given the important roles marketing and advertising have played at MCI, it is no surprise that this relationship is so solid. One might be led to ask which came first: Did the great ads stem from the great relationship or vice versa? There probably is no answer to that question. For whatever reason, serendipity perhaps, the personalities of both the people and the companies involved meshed from the beginning. Messner and his partners understand the MCI soul. And MCI, as Angela Dunlap has said, "always understood, and had a great respect for, the importance of advertising."

Young companies unsure about their own instincts will sometimes assume that the advertising agency knows best. And many times it does. But if MCI's experience teaches anything, it is that successful advertising is created when a company's agency

Action Steps

1. Interview ad agencies (even if your company already has one) to stir up the creative cauldron. Ask them the same kinds of questions you would ask propective employees: Do they understand your business? Can they give you new ideas to attain your goals?

2. Expand your discussions with your own agency beyond marketing talk. Help them to understand your long-range plans and business philosophy.

3. Review your agency's past work. Has it reflected the spirit of your company? Has it established in the marketplace the personality you were hoping to establish?

understands the company. If you feel that your advertising agency doesn't know your business or your company, find someone who does. Otherwise, all the money, all the clever words, and all the research will amount to nothing.

Tom Messner provides another insight into Bill McGowan's personality:

"[McGowan] loved talking about the [commercials] and showing them. And by the way, he loved awards, which is very funny to me, because the cliché in the ad business is that creative people—the art director, and the writer—salivate over these things that clients don't care about. Bill just loved awards.

"One year we won the American Marketing Association Award for effective advertising. I sent him a polite note saying that the dinner was on such and such a night. And he shows up. He and Orville [Wright] came and Jerry Taylor, and I thought that it was amazing that they would come to this stuff. They had all these awards in the case at his office."

Hit Back Fast

A key marketing tenet at MCI has been to answer AT&T's attack advertising swiftly and decisively. Without a creative rebuttal, negative images can quickly take a toll on both customers and employees.

Face It Head On

During the height of MCI's consumer long-distance telemarketing campaign, it was taking 100,000 customers away from AT&T every week. AT&T knew it had to stop this hemorrhaging so it introduced "Put It in Writing," which challenged MCI to present written proof that its services were cheaper.

This inventive approach sharply reduced the number of customers leaving AT&T by implying that MCI's telemarketers were dishonest nuisances who could be deflected by demanding proof. "Put It in Writing" demoralized MCI's telemarketers, stripping them of that certain buoyant quality that is so necessary in that line of work. It also gave a phrase to people to use to get the telemarketer off the phone, slowing down the sales process.

Jerry Taylor, then president of the Mid-Atlantic division, was called in to tackle the problem of rebuilding MCI's residential customer base. Taylor had proven himself as a creative marketer with the introductions of Execunet and MCI's residential service. He quickly realized that a deep-discount plan was needed to capture a portion of AT&T's 80 percent market share.

Although it was a fact that MCI's savings over AT&T were, at this point, no longer gigantic, Tom Messner, of Messner Vetere Berger McNamee Schmetterer Advertising Agency, suggested that MCI use an approach that politicians had used for years in campaign ads: Face the attacks head on. "Our advertising pitch to them (MCI) was that they had to stop being beaten up by AT&T," he said.

Of equal importance, in Price's view, is the uncomfortable position faced by consumers after such advertising appears. A corporate decision maker who chooses to go with MCI might feel threatened if the boss encounters unanswered attacks that call into question the employee's decision. Similarly, a residential consumer might feel uneasy around friends and family members who were selected for a calling circle if AT&T makes the product look like a bad deal.

Take the High Road

MCI recognized, however, that although it needed to answer AT&T in a strong manner, it didn't need to be negative in kind. The "Put It in Writing" spots came at the end of a bitter advertising war between AT&T, MCI, and Sprint, and the public was sick of the industry in-fighting. By devising Friends & Family, MCI responded to AT&T's charges while offering customers 20 percent savings on long-distance calls. But, equally important, Friends & Family let MCI take the high road—launching a feel-good advertising campaign in reaction to AT&T's mud-slinging accusations.

The plan also allowed telemarketers to become the good guys. They could get on the phone and say, "I am calling you at the request of your sister Sue who has joined our Friends & Family program. That means that she can call you long distance 20 percent more cheaply than she used to, effectively making her fifth telephone call free."

> By not responding, a company is tacitly agreeing with the negative things its competitor is saying.

Action Steps

1. Assemble a reconnaisance team that scans the media for new products and attacks (such as negative advertising) from your competition.

2. Assign responsibility to one person in sales, one in marketing, and one in advertising who meet regularly to discuss new competitors and attacks on your company and plan and execute counter measures.

3. Have your ad agency develop a few generic, ready-to-go ads that can be tailored as needed and produced quickly. Place them at the disposal of your response team.

Arm Your Employees With Detailed Ammunition

But MCI did not stop with Friends & Family. Later, under Price's direction, it also introduced Proof Positive, a product that further nullified the effectiveness of AT&T's allegations about MCI's charges and reliability. MCI had always put its savings in writing on its bills, but this new product took the proof a step further for business customers. The company's quarterly statements to customers actually compared the costs of each customer's long-distance calls using MCI to the cost of what identical calls would run using AT&T.

When AT&T under Joe Nacchio's leadership responded to Proof Positive with highly aggressive, attack-oriented ads, a different kind of advertising warfare came into being in what Bob Schmetterer considers to be another marketing breakthrough attributable to MCI. In this approach, MCI left no charge unanswered.

Likening Tim Price and Joe Nacchio to the opposing generals on a field of war, Schmetterer recalled it as a "brilliant" time. "They would run a spot, we would run another spot within a day. They would run a newspaper ad, we would run another the next day. We monitored with research how businesses were reacting and who was winning."

It was an extremely effective approach, particularly in the business arena, because it gave the sales force "air cover." Customer representatives and telemarketers had ready answers for any charges AT&T promulgated. Schmetterer believes this strategy was tremendously important internally and a big part of MCI's marketing success.

Living Lesson 17

Prepare for Success

One of the hardest problems to deal with for up-and-coming companies who are aggressively growing is the systems overload that occurs due to success.

Ready Your Operations

MCI first learned this lesson the hard way when it introduced Execunet, which resulted in a huge increase in business. Its entry into the residential market brought a similar experience. When MCI's initial ads aired in 1980, its phone banks were swamped beyond its wildest expectations, and the office ran out of order blanks within the first half-hour. Operators actually tried to keep up by writing on the walls when they ran out of paper and even in lipstick when the pens ran dry. MCI finally had no choice but to pull the ads and build up its staff to meet demand. "We were

> "We were like a mosquito that had tapped into an artery."
> – TIM PRICE

Action Steps

1. Before launching a new product, gather in the best estimates available on projected new sales volume.

2. Double those projected figures and determine what operations support will be needed such as extra telephone operators, telephone lines, order slips, and production.

3. Arrange for that extra support to be standing by. If the rush does not materialize, the extra support can always be canceled, but you may never recapture sales that are lost due to inadequate systems.

4. In the longer term, review your normal projected sales growth out five years. Determine whether you have planned for adequate support from operations to service that growth.

like a mosquito that had tapped into an artery," recalled Tim Price. "We'd obviously hit a gusher."

Make Sure All Orders Can Be Fulfilled

One of the best times to establish your reputation in people's minds is with quick fulfillment of their initial orders. Your image as a quality customer-service provider begins at this moment. When Friends & Family took off, MCI was ready with operators and customer-service operations to accommodate the substantial increase in business.

MCI's ability to capitalize on strong demand for Friends & Family and to support it with high-quality customer service helped it take three points of market share from AT&T within the first two years of its offering. The remarkably successful program, which had a million customers within three months of its launch and was in ten million households in a little over eighteen months, today is MCI's flagship offering in long-distance telephone services.

Do Not Trust Conventional Wisdom

Time and time again it is the new idea that captures the hearts and minds of the consumer. You must be constantly on the lookout for new needs and new ways to fill them.

Question Assumptions

MCI learned through experience to keep industry forecasts and opinion polls in proper perspective. Louis Harris, a respected pollster, advised MCI not to aim its products at people over fifty years of age. "Write off the fifty and above crowd," he said.

That's exactly what MCI did for years. Before Friends & Family, the majority of MCI's customers were under forty, a fact that management attributed to older people not identifying with the company's feisty image. But after MCI introduced the new product, with its decidedly friendlier image, older people began signing up.

Friends & Family helped soften MCI's image and increased its appeal to an older audience, many of whom often call loved ones far away. Grandparents, in particular, jumped at the opportunity to speak with their grandchildren more frequently.

Target the Impossible

There are often simple solutions to seemingly impossible problems. But you must actively look for them in sometimes very out-of-the-way places to find them. Consider MCI's huge success with 1-800-COLLECT. AT&T owned the $3 billion-a-year collect-call market until fairly recently. It was a foregone conclusion among AT&T's competitors that collect callers didn't care about price since they weren't the ones footing the bill. The giant, they believed, had a lock. But in 1993, Taylor began questioning this assumption—if MCI could grab just a small piece of that huge market, he reasoned, it would be a huge revenue gain.

To that end, MCI began conducting research on exactly who made the most collect calls. It found that of the 300 million made annually, 24 percent are made by military personnel calling home, and another 33 percent are children calling home. It also discovered that 70 percent of all collect callers are under age thirty. While there was no evidence to suggest that the callers were concerned about price, MCI guessed that family members could prevail upon loved ones to use a less expensive service if one were available.

Traditionally, collect calls were among the most expensive to make. Depending on where and when the call was made, it could cost twice as much as a direct-dialed call. One reason was that nearly 60 percent of collect calls originated from pay phones that had long-term contracts with AT&T or other carriers.

MCI's idea was that by allowing callers to connect directly with an MCI operator, MCI could bypass the carriers that contract with pay-phone owners. Then it could send the call through and bill the regional phone company just as AT&T did. This way, MCI could offer significant savings to customers—up to 44 percent off AT&T's collect-call rates.

> My brother, who is a priest, always cringes [when I say that] I've always believed that although the meek may inherit the earth, they won't increase market share.
>
> – BILL MCGOWAN

The opportunity was irresistible, and MCI decided to move. It initiated and launched 1-800-COLLECT within seventy days—warp speed for most companies, but business as usual for MCI.

Just like MCI's Friends & Family service, which Taylor also masterminded, 1-800-COLLECT had a personality. Where MCI's advertising positioned Friends & Family as warm and friendly with appeal for an older market and women, 1-800-COLLECT was youthful and "in."

"The ads gave a sense that you were doing something different," says Price. "Dialing zero was the old-fashioned way. You were behind the times if you did it. With 1-800-COLLECT, you were hip." The ad campaign included television spots featuring one-time David Letterman sidekick, Larry "Bud" Melman, and biplanes that flew over beaches carrying the 1-800-COLLECT logo.

MCI also advertised the product without mentioning the corporate name. This was no accident of omission: the company reasoned that identifying itself might lead people to assume that only MCI customers could use the service when, of course, anyone could. In choosing to use 1-800-COLLECT, many regular AT&T customers didn't realize they were patronizing MCI.

Today, 1-800-COLLECT owns 30 percent of the collect-call market and provides MCI with more than $300 million in annual revenue, with very high margins. It has no costs other than for advertising and marketing, which is handled by three MCI employees.

Even AT&T had to concede MCI's coup. "They moved aggressively and quickly in a part of the market that had not been focused on," Jack McMaster, AT&T's director of consumer-product marketing, told *Newsweek* magazine.

MCI's success with 1-800 COLLECT is a perfect example of how important it is not to accept things as they are—you must never assume that because there are no competitors in a product category, there can be no competitors. It's easy to say, "If it could be done, someone would be doing it." But it is much harder to do the market research, figure out an angle that has not been explored, and come up with an effective advertising plan to exploit it. As MCI has repeatedly shown, making such an effort is almost always worth it.

Action Steps

1. List the industry limitations that are commonly accepted by colleagues and experts in your field of work such as limits of sales volume, market share, expansion, market saturation, regulatory constraints, national boundaries—everything you can think of.

2. Exercise your skepticism by questioning the underlying assumptions of these limitations. Find one that shows potential for overcoming.

3. Persevere on breaking open this one limitation. Have everyone in your division or company work on a solution. Talk about it to your friends, colleagues, people at cocktail parties and conferences, taxi drivers, web surfers, the passenger next to you on the airplane—anyone and everyone—because solutions often come from strange sources. Don't give up till you've cracked the code.

Living Lesson 19

Make the Most of Hard Times

All savvy financial analysts will tell you that hard times are also times of great opportunity. This is especially true for flexible, up-and-coming companies who have the ability to offer real value to consumers looking to save money.

Help Your Customers Save Money

In 1974, for example, when MCI introduced Execunet (its shared private-line long-distance service), the country was in a recession. It was precisely this depressed economy that enabled MCI to make its first major move against the world's biggest monopoly. Its weapon was Execunet.

During a downturn, businesses look for ways to cut spending, and Execunet offered them a way to do just that. Even though switching to MCI from AT&T certainly was considered a bold move at the time, managers at companies that are in distress are more likely to take such action to stem the flow of red ink. They are willing to shake up the status quo because jobs are on the line.

When MCI approached General Motors, the car manufacturer was faced with having to dramatically cut their budget. "The

Action Steps

1. Review the list of your company's products and services.
2. For each product list at least three ways it can save your customers money in both the short and long term.
3. Experiment by incorporating these points into a few advertising placements and gauge customer response.

guy's choice was to go with MCI or to cut people," explained Jerry Taylor. GM wisely chose to spare the employees.

Time Your Product Launch

MCI's success with Execunet shows that there is a right time to launch a product. For Execunet that right time was smack in the middle of a recession. If MCI had launched Execunet in boom times, it would have squandered a good deal of Execunet's appeal to cost consciousness. It would not have enjoyed such initial success without this vital hook to help it break through entrenched complacency. On the other hand, a more expensive product that offers a lot of bells and whistles will do well only if it is launched in boom times.

If your company offers a product or service that saves consumers money, sailing into the wind could be your best option. Then, by the time the economy has changed direction, your customers will be convinced of the quality of your product, and you will be firmly established in the marketplace.

Give Customers What They Want

Living Lesson 20

The strategy seems simple: Ask consumers what they want and then give it to them. But it's amazing how often companies fail to do that.

Use Focus Groups

> The RBOCs don't have a clue about customer service, because they've never needed to have it. If you have all the customers, what do you need customer service for?
>
> – BOB SCHMETTERER

MCI's story would be a very different one if it had not acquiesced to suggestions by its advertising agency, Ally & Gargano, to conduct focus groups before it launched a test of its residential services. Without focus groups, MCI wouldn't have been able to determine customer preferences and usage. Because AT&T had been the only long-distance carrier, data on customer preferences was unavailable. (AT&T had the information, of course, but it wasn't about to share it with MCI.)

MCI's inclination was to apply what it knew about the business market to the residential market. It thought about charging a minimum monthly fee, similar to the $75 fee it charged its business clients. Yet Ally & Gargano persisted, arguing that most residential customers didn't even know AT&T had competition. If MCI wanted to best serve residential customers, the agency con-

Action Steps

1. Distinguish the primary needs (such as need for food, transportation, communication, entertainment, information processing) from the secondary needs (such as the need for status, economy, indulgence, acceptance, affection) that each of your products fulfills.

2. List some secondary needs not addressed by your products.

3. Develop at least one new product that will address one of these customer needs that your products currently do not satisfy.

tended, it needed to find out firsthand how it could make long-distance service attractive to them.

To MCI's surprise, residential customers weren't like business customers at all. The focus groups revealed that they would be unwilling to pay a regular monthly fee unless they were certain of spending that amount every month. This new information led MCI to keep the charge but lower it to $10, a sum that virtually every household routinely spent on monthly long-distance service. (Incidentally, MCI also decided to drop its business account minimum to $10 and was rewarded with a flurry of new customers.)

The groups also showed that consumers blamed themselves, not the phone company, for the high cost of their bills. That revelation triggered the idea that became the cornerstone of MCI's advertising position, and one that is still in evidence today: "You're not talking too much; you're just paying too much."

Do Not Project Your Own Needs and Desires on the Consumer

Although MCI has successfully used focus groups throughout its history, it has occasionally lost sight of market realities in the glare of exciting technological advances. One such case is MCI Mail. Introduced in 1983 as a combination e-mail and courier service, it was billed as the "nation's new postal system." MCI spent $30 million on mass-market television and newspaper ads the first year. The only problem was that, at the time, most residences didn't have personal computers, and fewer still had the modems needed to connect to the electronic mail service. Even in the business world, executives were, for the most part computer illiterate and reluctant to move away from traditional courier services in handling time-sensitive documents. In the long run, MCI Mail did help position the company as a forward-thinking organization, but as an individual product, MCI Mail failed because consumers had no use for it at the time.

Although an essential part of any business's success is its ability to anticipate consumer needs and desires, companies must be vigilant about differentiating between what consumers want and what the company thinks they should want.

> Bell never surveys customers. Bell offers you what it is convenient for Bell to offer you. And if you don't like it, then try Dixie cups for awhile.
>
> – BILL MCGOWAN

Sell to the Big Guys

There is a marked difference in the approach you must use to sell to large companies as opposed to small ones. MCI, for most of its early years, sold to small companies, and when it decided to make a concentrated effort to sell to larger accounts, it had to change the way it did business.

Hire an Experienced Mentor

To start your campaign of selling to large companies, put a person with experience in charge of the program. He or she will work as the mentor or coach coordinating the sales teams on the front lines. Carl Vorder Bruegge, who was MCI's senior vice president of sales before he retired in 1986, was an experienced salesman with extensive experience in computer and financial services when he was hired in 1972. His background included fourteen years in sales and marketing at IBM, then a shorter stay in marketing with Leasco Corporation, where he was responsible for computer time-sharing, among other things. When he was hired at MCI, he immediately went after large accounts along the company's Chicago-to-St. Louis route. He hired ten teams of three persons each: a salesperson, a technical person, and a junior salesperson. Each team was devoted to specific industries—such as banking, insurance, aerospace, transportation, and so forth—and had enough experience in its designated industries to understand their communications needs.

Their efforts paid off when MCI for the first time in its young life captured a number of large accounts, a remarkable achievement for such a small company. (Most larger companies are more comfortable dealing with larger vendors simply because larger vendors have more resources to assist their customers with.)

Set Up Sales Teams

Each sales team should have one person with expertise in the specific industry targeted, a seasoned sales pro, and one or more young, energetic junior salesperson. (Graduating college students provide an excellent pool of raw talent for the latter category.) The leader mentor should indoctrinate these sales teams in a concerted and coordinated sales effort. They should be instructed in how to dress, talk, and behave when selling to large-account middle managers. Most importantly, their message should be focused and well polished.

Pitch to Emotions

Where possible, craft your presentation to play on the emotions of businesspeople. MCI, for instance, played up the necessity of competition (a strong, inherent belief of almost all American businesspeople) and the ability to choose from a pool of vendors (most businesspeople prefer to have more than one vendor from which to choose).

Get Your Foot in the Door

Most large accounts are reluctant to hand over a big chunk of their business to relatively unknown or untried companies. It is

Action Steps

1. Read your target company's mission statement and any recent comments and statements in the media by top executives on the company's purpose and ambitions.

2. List all the goals that the company is working toward as well as the philosophies and emotions that underlie these goals.

3. List all of the ways that your product or service can help the potential customer toward those goals and incorporate these points clearly into your presentations and overall pitch.

essential, therefore, that you gain a foothold at the new account any way you can. MCI's solution, as it quickly recognized that it was not making great gains, was to ask for trial orders. After the company had a chance to prove itself, its salespeople would ask for all of the large account's communications business. MCI almost always made progress with this approach, which proved much more successful than going for all or nothing. Another plausible way to gain credibility is to win the account of a small division of the larger company you are targeting. Prove yourself there and you will have allies inside to help you make the big sale.

Partner Up With a Major Player

In the early 1980s Bill McGowan recognized that MCI had let its large-account business lapse. He knew that the company had relatively few salespeople who could sell to the likes of a Johnson & Johnson and that MCI needed a jump-start if it were to cater to large accounts.

That was one of the key factors that led to the 1985 alliance with IBM. McGowan was trying once again to move from A to Z along the shortest possible route. And it worked. The credibility MCI gained from its association with IBM helped it reposition itself in the large-account market.

In addition, MCI's salespeople got to accompany IBM's crew on calls to large accounts. The training MCI salespeople received from one of the marketing leaders of American industry proved invaluable.

Chapter 11

Financial Strategies

MCI is the company that should never have been...and was inevitable all along.

—BILL MCGOWAN

Living Lesson 22

Educate Yourself

Financing is an art that requires a broad knowledge far beyond the numbers. To create optimum value for a company and investors, you need an understanding of your business and its industry, government regulations, technology, the financial markets, and social and economic factors. Financing never takes place in a vacuum. And although it's tempting to focus only on immediate factors, the most effective offering will have a far broader perspective.

Know Your Company

The first step is understanding the nature of your own company and its sphere of influence by having a firm grasp on your future plans, products, customers, and suppliers. This advice may seem puzzling because you probably feel you already understand your own company. But everyone, even the Fortune 1000, finds it difficult to complete a critical analysis of themselves that is accurate and objective. Management is often so close to the situation that they have trouble gaining real perspective, and must work to overcome their insider bias. An inaccurate, subjective analysis can lead to serious missteps in the market place.

A truly successful financing does more than bring you money—it furthers the goals and ambitions of the company. For example, if you expect to grow in ways that will delay cash flow for a few years, your financings should take this into account. You may arrange to pay a little more interest so you may issue longer-term instruments, for example. If you anticipate a cash crunch, on the other hand, you might focus on equity rather than debt financings, so you may avoid cash interest payments.

Mike Milken illustrates the multidimensional approach to financing with the Milken Cube, the six sides of which are reproduced on pages 303–306. Each side represents one of the perspectives key to a successful financing strategy.

Financial Strategies

Know Your Industry

The next step is to study your industry both nationally and internationally. Identify the industry's level of maturity and how your competitors fit into its framework. What are their positions in the marketplace in relation to your business? How will their growth, or lack thereof, affect you? Part of the reason for MCI's successful 1983 offering was its position as a cutting-edge company in an expanding industry. Its major competition, meanwhile, was big and bureaucratic, having been around literally since the invention of the telephone.

An important industry-related component is the effect of technological advances. Capital-intensive industries, such as telecommunications, not only need a vast amount of equipment to generate their products, but also must upgrade and replace that equipment as technological advances are made. Had it not understood this fact about its industry, MCI could not have accurately forecast its capital requirements.

Know the Regulatory Environment

Another crucial consideration is the effect that government regulations have on one's company, industry, and markets—both consumer and financial. (See Living Lessons 9 and 10, which deal with regulatory strategy.) The telecommunications industry illustrates better than most that a company cannot operate in a vacuum. Vigilance in pursuing positive legislation is one reason MCI survived at all. Victories in this area also helped to promote MCI as a viable contender in the eyes of the capital markets.

Know the Capital Markets

In addition to knowing the hard facts about your company and the factors that influence it, you must study the state of the financial and capital markets to float a successful offering. Success comes only to those who are adept at gauging such issues as supply and demand of offerings, general market trends, and industry and company appeal to investors.

Consider the example of Wired Ventures Inc.'s two attempts to go public in 1996, first in July and later in October. Wired had made a splash with the super-hip *Wired* magazine, which covered the exploding Internet industry with an appropriate creative style, but had yet to turn a profit. Still, the company hoped to cash in on the Internet enthusiasm on Wall Street that was pushing software companies like Netscape to skyrocketing share prices. But the company misread the overall market mood in  important ways. The first IPO attempt was pulled after a downturn in Internet stocks made for a dismal investor reception. The company cut its price and valuation for the second offering, but still failed to convince investors. Many believed that Wired's failure to understand its own image in the financial market was a critical problem—while the company visualized itself as a high-growth Internet company (with high price-to-earnings multiples and valuations), many investors saw it simply as a magazine publisher.

Pay Attention to Trends in Society

Although some companies focus strictly on hard economic data, such as the direction of interest rates, the social environment plays an equally important role in assessing your financing needs. Does your product or service mesh with the mood of the country? The fitness

Financial Strategies

movement is a good example of an important change in national perceptions: The market for health clubs, exercise equipment, and other "healthy lifestyle" products and services has exploded in recent years.

You will need the knowledge gathered from these different arenas to accurately assess your financing options. Looking at MCI's 1983 offering, for example, it's apparent that the aforementioned considerations played a role in the decision-making

process. MCI knew—had always known—that it stood in the shadow of a giant. The fight was colossal and so, too, was the amount of cash the company needed just to stay in the ring. MCI was using capital at the rate of $50 million a month in the early 1980s. Its voracious appetite reflected not only the nature of the competition, but also the very nature of the industry— one that technology was rapidly transforming.

The government was helping to fuel the industry's growth as was the investment community. Ronald Reagan was President, and his administration favored competition both at home and abroad, seeking to relax regulation in the United States while promoting free trade outside its borders. The capital markets were more receptive to new companies regardless of size or credit rating than they had ever been. Interest rates, though still relatively high (the prime rate was about 11 percent in 1983), had declined by 25 percent in the previous two years. And the fact that the spread between U.S. government and non-investment-grade debt was at its narrowest level in more than a decade was a clear indicator that the time was right to issue debt: The risk-reward trade off of high yield (junk) bonds made them an attractive alternative to investing in Uncle Sam.

Fortunately, MCI was in a strong position to take advantage of the receptive marketplace. Its stock had risen sharply—from well under $5 a share in 1981 to more than $12.25 at the end of June. But most importantly, MCI had the landmark AT&T antitrust victory in its corner.

The key to effective financing is the ability to interpret the meaning of these various factors when fashioning a strategy. Just because the bond-warrant issue was successful for MCI in 1983 didn't mean that it would always be the best financing alternative. MCI was adept at reading the marketplace and incorporating that assessment into its choice of financing instruments.

In the early 1980s, such companies as McCaw Cellular, Turner Broadcasting, and Viacom International forsaw huge new markets for services like cable television and cellular phones, but knew that cash flow problems would be significant. These were markets that would require significant up-front investment but would yield delayed profits.

All three companies understood this cash flow dilemma well, and made appropriate adjustments in their financing plans. McCaw issued twenty-year discounted convertible debentures, which paid no cash interest for five years and allowed his company to retain its much needed cash flow. Turner Broadcasting turned to zero-coupon notes, which deferred interest payment until maturity, while Viacom used pay-in-kind securities, which paid interest only in the form of additional securities for five years after issuance.

Learn the Tools of the Trade

MCI had several financial instruments at its disposal when it began to build its capital structure. In today's marketplace, an even broader array of financial instruments is available to the knowledgeable financier.

1. Common stock. Common stock is a share of equity in a company. A shareholder owns a portion of the company equivalent to the number of shares held times the price per share. Share ownership bestows the right to vote on important matters such as electing the company's board of directors, thereby giving shareholders the power to indirectly control the company. Shareholders benefit from the company's success through stock price appreciation and often through dividend payments. Common stockholders have only a residual claim to the company's assets behind creditors, bondholders, the IRS, and preferred shareholders in the event the company files for bankruptcy.

2. Preferred stock. As the name suggests, preferred stock has certain preferential rights. These shares of equity carry guaranteed dividends at a fixed rate (while common stock dividends are strictly at the discretion of management). The dividends are generally cumulative, meaning that if they are not paid for any reason, they accumulate and must be paid in full before common stockholders are paid anything. Some preferred stock is convertible into a predetermined number of common shares at a fixed price. Preferred stock is a more beneficial capital-raising instrument for non-tax-paying entities because the dividends paid out do not qualify as a tax-deductible expense (unlike the interest on debt). Additionally, corporate investors often find them attractive because many preferred stocks offer 80 percent tax-deductible dividends.

3. Subordinated debt. Subordinated debt is unsecured debt obligation that is junior in liquidation preference to senior secured and secured debt. Because subordinated debt is further down in the capitalization ladder, it carries more risk than senior debt. This means that the interest rate, or yield, that a company must pay on subordinated debt is usually more expensive.

4. Senior debt. Senior debt is debt obligation that ranks below secured debt in event of default or liquidation but above subordinated debt and preferred and common stock. Senior debt usually includes loans from banks or insurance companies and any bonds or notes not clearly designated as junior or subordinated.

Senior debt has a number of variations. It may be fixed or floating rate, convertible or straight, long or short in maturity, or issued by a public or private company. It may pay its holders quarterly, semi-annually, or annually, and it may have features like a call, sinking fund, and special covenants. Because of its senior ranking, it may have more restrictive covenants than subordinated debt issues.

5. Convertible bonds. Convertible bonds are bonds that can be exchanged for a specified number of shares of common stock in the future, usually at a predetermined price. A convertible bond is typically unsecured debt and would rank above preferred and common stock in the event of a bankruptcy. In addition to scheduled interest payments, equity participation is offered to the investor if the stock price rises above the conversion price. Convertible bonds are booked as debt on the balance sheet unless they are converted to equity. Because of the "equity kicker," the coupon rate is typically lower than that of a comparable straight bond. Although convertible bonds can be short term, they are typically long-term obligations (i.e., fifteen years or more).

A simplified example: Say your company wants to obtain financing through a bond issue. Your banker tells you that your likely interest rate for a twenty-year issue would be 10 percent. You are reluctant to pay that much interest and ask them to explore convertible bonds. Your share price is currently at $20, but is believed to have good upside potential. Based on that assessment, your banker tells you that with a convertible bond, offering investors the chance to convert at $25 per share, your interest rate would be only 8 percent. The advantages to the issuing company are a lower interest rate than a straight bond issue, and a higher price for stock than a straight equity issue. Key disadvantages include equity dilution if investors do convert and a lower stock price than you might have been able to get later.

Action Steps

1. Create a database of information about your company and industry. Include newspaper and magazine articles, analyst reports, and annual reports for publicly traded companies.

2. Study their capital structures and create a shorthand financial portrait for these companies and your company.

3. Examine how the market perceives your industry by tracking industry stock performance and reading analyst reports. Contact at least one brokerage analyst who covers your industry and establish a relationship. They will be happy for another information source, and you will gain valuable insight into Wall Street's perspective on your business. Offer to attend or speak at industry conferences.

6. Combination bond-warrant units. Combination bond-warrant units are convertible securities comprised of a bond and a set number of equity warrants (options). In accounting for these units, the issuer records the bond as debt and the warrants as equity. (In contrast, conventional convertible bonds are recorded entirely as debt until converted.) If desired, the bond portion is usable as currency—generally at par—to exercise the warrants. Unlike conventional convertible securities, the debt (bond) and equity (warrants) can be traded separately.

7. Vendor financing. Vendor financing is credit extended to a company by its suppliers. When approaching a vendor for a loan, the key is to make sure that the vendor is cash-rich and a source of cheap capital—the vendor's cost of capital should be lower than your company's. Vendor financing is particularly effective when a company has numerous suppliers. The suppliers may be forced to compete not only on the products and services they provide but also on the credit terms they extend. Vendors have a vested interest in keeping their customers healthy, but such financing is complex to negotiate and can give vendors too much control over a company. Thus, this type of financing should be used only when no other affordable options are available.

Build and Balance Your Capital Structure

Living Lesson 23

Suppose you were constructing a plan to finance your company. The first question you should ask is about your desired capital structure—the combination of debt and equity on a company's balance sheet. The simplest way to think of capital structure is as drawers in a safe that you are filling up.

One of the more important concepts behind capital structure is that of stakeholder priority—that is the precedence of claims on a company's assets in the event of bankruptcy. This idea can be illustrated as drawers in a safe, which can be opened only in descending order from the top. In this example, senior secured debt holders will be paid first, then the senior debt holders, and so on. Equity holders are always at the bottom of the priority scale, with preferred stockholders (as the name might imply) taking precedence over common stockholders.

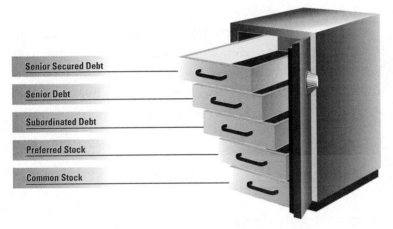

Senior Secured Debt
Senior Debt
Subordinated Debt
Preferred Stock
Common Stock

Imagine that the drawers of the safe can be opened in descending order only. In the event of a bankruptcy, those with access to the top drawers will be paid first. Thus the top drawers represent senior secured debt; they provide the most security for their holders. The bottom drawers would represent preferred and common stock, and provide the least security because their holders are the last to get paid in the event of bankruptcy. Because the securities at the top are considered less risky, they appeal to a wider range of investors and thus have the advantage of being easier to sell in an unfavorable market.

When the market is receptive to providing capital through both equity and debt offerings, a corporate manager should utilize the most junior levels, that is, fill up the bottom drawers. Once the bottom layers are in place, various layers of debt can be added.

Financial Strategies

The middle layers, known as mezzanine financing, should be used when a company is in relative favor in the marketplace. First come subordinated debentures. These securities have no assets behind them to support the value of the bonds and thus are often more difficult to sell. For that reason, a solid layer of subordinated debt should be constructed when the opportunity arises. The next layer is unsecured senior debt. It is not tied to any assets but has priority in claims over other securities in the event of liquidation.

In the top drawer and last to be placed is the senior secured debt. This type of debt should be saved for when it's most needed, that is, when times are tough and market conditions are less favorable.

Like many upstart companies, MCI financed its initial growth in the private markets with senior secured debt, namely vendor financing and bank loans. For a fledgling company trying to make a start in a recessionary economy, it was the most accessible form of capital, but also the most expensive.

Action Steps

1. Diagram your capital structure as it stands, listing the amount of a particular instrument and the percentage of the capital structure it represents. (For smaller businesses, this list would include such financing sources as bank loans, personal loans, and so forth.)

2. Rank the financial instruments by the percentage of the capital structure each accounts for, lowest percentage on top to highest percentage at the bottom.

3. Map out your optimal capital structure, taking into account your anticipated needs for flexibility, cash flow, and so forth. For example, if you anticipate limited cash flow for one to two years your optimum structure could be skewed toward common equity, which requires no interest payments. Your goal might be 70 percent equity and 30 percent debt.

4. Compare your capital structure to the paradigm you've delineated in step 3. What steps could you take to get closer to that model?

In 1972 MCI entered the public equity markets at the insistence of its bank lenders. The banks would only agree to lend MCI additional funds if the company floated a public offering and attained a market valuation of more than $100 million. That initial public offering marked the start of a twenty-four-year effort by MCI to attain the optimal capital structure via the public markets.

The company started by selling common stock, then preferred stock, then subordinated debentures, and it continued to issue financial instruments that built its capital structure in a logical order. Even today MCI actively works toward building and bettering its capital structure to fully fit its needs. In 1996, for example, MCI reduced its debt to less than 35 percent of its total capitalization in an effort to prepare for a future made more uncertain by the Telecommunications Act.

Partner With Vendors

Vendors are often in a better position than banks to make decisions about helping a customer because they understand your industry and the strengths and weaknesses of the players. They may be more adept at knowing which customers are going to be able to pay them back and which aren't. Getting help from a vendor is like a vote of confidence. Be aware, however, that relying too heavily on vendor resources can cause your company to lose negotiating power in the future.

> There were two kinds of vendors: those who were independent and those owned by AT&T. The ones owned by AT&T wouldn't sell a competitor equipment under any circumstances.
>
> – BILL MCGOWAN

Purchase Products and Services on Vendor Credit

Whereas banks will sometimes shy away from lending to customers in a single industry because they want to spread their risk, vendors often are tied to just one sector anyway and will do everything they can to help their customers succeed. Vendors recognize that if the borrowing company grows, so too will the market for their goods and services. Vendors also worry, of course, about spreading their risk, but they do so by financing a number of companies in the industry so that if one company goes under, most of the money they've loaned is not affected. In the event of a bankruptcy, a vendor is in a better position to make use of (or sell) a company's assets, such as equipment and factory space. Vendors, therefore, will usually be more flexible on continuing your line of credit while cutting special deals and offering easy pay-back schedules to help keep your company afloat. Remember, they have a vested interest in your corporate health and well being.

Secure Loan Guarantees by Leveraging Off Vendor Faith

Another way to utilize the financial strength of vendors is to arrange for them to provide loan guarantees to reluctant banks.

This strategy will usually be expensive in the long run, but it is a great way to avoid calamity when a company finds itself in a short-term bind for cash. In its early years, for example, MCI did not have the deep pockets needed to finance construction of a nationwide telecommunications network. In 1972 with the company starved for capital, Stan Scheinman, MCI's chief financial officer, went to the company's vendors for loan guarantees that enabled MCI to get a $64 million line of credit from its bankers. In return for the loan guarantees, MCI granted the vendors the ability to raise their equipment prices by up to 50 percent—a hefty price to pay but probably the best deal they could negotiate at the time. The vendors gain by preserving another client who will purchase their products and services and who will in all likelihood be loyal and indebted to them. Their loan guarantees become an investment in the future.

> Vendors recognize that if the borrowing company grows, so too will the market for their goods and services.

Action Steps

1. List your vendors and rank them in order of the amount of business you represent to them over various projected time periods (i.e., one year, five years, ten years).

2. Assess your agreements and pricing structures with these numbers in mind. Are you being treated fairly in light of your potential importance to your vendors? If warranted, initiate renegotiation with these numbers in hand. Review pricing, payment terms, and financing support. Offer long-term contracts or other trade-offs (e.g., higher prices in exchange for loan guarantees) that may address the immediate needs of your company.

3. Draw up a plan that would replace at least one financing source (say, a bank loan) with vendor financing. Consider seriously the costs and benefits.

4. Ask your bank about the possibility of having your loans being guaranteed by one of your vendors. Would this lower your interest rate? If so, explore the possibility with the vendors who are at the top of your list in step 1.

This approach can be difficult, of course, if vendors feel beholden to one big customer. When MCI first appealed for help, AT&T was the largest customer for most of the industry vendors. And, understandably, some had concerns that any effort to keep MCI afloat would jeopardize their relationship with AT&T. Even those who were independent and sold large equipment orders to AT&T were afraid their contracts would be jeopardized if they sold to a competitor.

Find Vendors That Are New to the Business

MCI solved this problem by approaching vendors that were upstarts themselves. Such companies were more willing to risk their relationship with AT&T to help build MCI and the industry—and thus their own fortunes. By supporting MCI at a critical time in its growth, these vendors gained MCI's loyalty and lucrative business for years to come.

Living Lesson 25

Be Ready to Go Public

Although most companies enter the public market of their own volition, some are pushed into it by bankers and venture capitalists. Such situations occur again and again in the financial marketplace. Venture capitalists have funded the growth of many startup companies, including America Online, Netscape Communications, and Sun Microsystems, in exchange for ownership rights. Once venture capitalists decide that, based on market conditions, the time is right, they encourage startups to go public.

If your company is a candidate for a public offering in the foreseeable future, do your homework now. It doesn't cost any money to be prepared—to know that your capital structure is well balanced, your cash flow is healthy, your books are in order, you've researched investment bankers, and analyzed the mood of the financial markets. In this way, you will be able to time your offering properly and reap all the benefits a public offering can generate for a company.

And a successful offering provides advantages beyond the obvious cash infusion. The benefits of a successful IPO include:

- Non-interest-bearing capital to support growth, increase working capital, and invest in plant and equipment.

- Enhanced ability to borrow because debt-to-equity ratio is improved.

- Increased liquidity, giving investors an exit strategy.

- Greater ease in valuing the company.

- Enhanced prestige from being listed on the stock market—potentially useful with customers, suppliers, and employees.

- Ability to raise more equity in the after-market with lower-cost shelf registrations, which facilitate more rapid securities sales in the future.

> Investors in 1996 were generally most receptive to IPOs from companies with net earnings in excess of $1 million, current revenues in the range of $20 million, and the potential to achieve revenues of $100 million in five years.

Financial Strategies

Of course, there are drawbacks to going public as well. They include:

- Increased pressure for short-term performance from shareholders and analysts.

- Increased disclosure responsibilities from regulatory agencies (ownership disclosure, quarterly and annual reports, and so forth).

- Increased disclosure pressure from the shareholder community beyond regulatory requirements. (For example, many analysts believe companies should release pricing information as a bare minimum measure of the company's health, while many companies believe pricing to be highly sensitive competitive information.)

McGowan and Scheinman had begun discussions about an IPO for MCI in 1971. They realized that the series of private investment deals MCI had relied on up to that time were no longer feasible and that the company needed a strong financial base. The logical thing was to sell stock to the public. When the banking group led by First National Bank of Chicago made floating a successful IPO one of the prerequisites for obtaining the $64 million line of credit, MCI quickly put its books in order and was ready to go.

Action Steps

1. Make a checklist of the things that will need to be in place for you to go public including, but not limited to, sound capital structure, healthy cash flow, orderly bookkeeping, analysis of market trends, securities research on your industry, and a good investment banker.

2. Develop a plan of improvement or execution for each item on your list. Where possible, delegate the work to a competent colleague to keep things moving.

3. Set a deadline for each item on your list and mobilize people behind the exciting prospect of going public. (Even if you choose not to go public, this process will leave your company better organized and ready for other options.)

Living Lesson 26

Time Your Offering

There are a number of important variables to consider when entering the public financing markets, but timing may well be the most critical. No matter how strong your company's story, if your offering catches the market at the wrong time, it will face a hard sell at best.

Obviously, waiting for the best time for an offering is a luxury many companies do not have.

If an offering is made at the wrong time, not only will your company raise less money, it may do so at much higher interest rates or lower stock prices. And who knows when the best *time* will be? For most companies, this issue comes down to avoiding blatantly bad markets.

Timing was very much an issue when a cash crunch forced MCI into another public stock offering in 1975, and the company learned a hard lesson. The market was depressed, MCI's sales and installation objectives were not on target, and AT&T had asked a federal appeals court to prohibit the expansion of Execunet, MCI's only profit center. Its underwriter at the time, Allen & Company, finding that MCI's precarious position was

> MCI made its initial public offering (IPO) of common stock on June 22, 1972, when it issued slightly over three million common shares at $10 apiece. Net proceeds amounted to $30.2 million.

Action Steps

1. Evaluate the financial offerings floated by companies in your industry during the last three years. (For example, a review by MCI in 1983 of similar offerings would have looked like the table on the following page.)

2. Track the valuations using standard benchmarks such as price-to-earnings or price-to-sales. Determine whether the trend is generally favorable or hostile to new offerings.

Offerings Completed in First Quarter of 1983

Date	Company	Security Type	Amount ($m)	Coupon (%)	Maturity	Rating	YTM (%)
1/5/83	Crystal Oil	Subordinated debentures	$80	12 5/8	12/15/01	NR/NR	16.50
1/12/83	Pacific Telecom	Common shares	$27	NA	NA	NA	NA
1/17/83	City Investing	Subordinated debentures	$50	13 1/2	2/1/03	Ba1/BB	13.65
1/24/83	Jim Walter	Subordinated debentures	$50	13 1/8	2/1/93	Ba2/BB-	13.13
1/24/83	Jim Walter	Subordinated debentures	$100	13 3/4	12/1/03	Ba2/BB-	13.75
2/2/83	U.S. Telephone	Common shares	$20	NA	NA	NA	NA
2/3/83	Oppenheimer	Subordinated debentures	$30	12 3/4	2/1/03	NR/BB	14.25
2/25/83	Rohr Industries	Subordinated debentures	$25	13 1/2	3/1/98	B2/B+	13.65
3/8/83	Tesoro Petroleum	Subordinated debentures	$120	12 3/4	3/15/01	Ba3/B+	15.29
3/31/83	Mobile Communications	Common shares	$15	NA	NA	NA	NA

dampening investment interest, repeatedly delayed a planned offering of common stock and warrants, hoping MCI's stock price would break out of the $2-per-share doldrums where it seemed to be stuck.

But with MCI running out of cash (and options), the company finally had no choice but to come to market with an offering that brought in significantly less than the $15 million originally envisioned. In late 1975 MCI sold 1.2 million stock-warrant units at $8 apiece, netting $8.5 million with the potential for $12 million more if all the warrants were exercised. In the process MCI gave up nearly 10 million shares, or about 40 percent of its equity.

Never Run Out of Cash

Living Lesson 27

As Harold Geneen, former chairman of the hospitality and entertainment conglomerate ITT Corporation, is fond of saying: "There's only one mistake in business; everything else you can recover from. If you run out of cash, they take you out of the game."

Cash flow is one of the most important financial considerations of any company. It is the life blood of the corporate body and its unrestricted movement not only will enable you to time your offerings properly, it is essential to the very survival of the company itself.

Bill McGowan, who always hated the idea of selling equity, was especially perturbed over the 1975 stock-warrant offering, which he considered to be the worst deal of his life. However, the company had let itself get backed into a corner because it

Cash Flow Projections: A Simple Example

	2000	2001	2002	2003	2004	2005
Funds from operations	($200)	($100)	$0	$50	$100	$200
Funds from investing	($200)	($50)	($50)	($50)	($50)	($50)
Funds from financing	$200	$0	$0	$0	$0	$0
Change in cash flow	**($200)**	**($150)**	**($50)**	**$0**	**$50**	**$150**
Beginning cash	$500	$300	$150	$100	$100	$150
Ending cash	$300	$150	$100	$100	$150	$300

In this simple example, we assume that a business loses money from operations in its first three years, then begins to show a profit. We also assume a large initial investment of $200,000, followed up by yearly upkeep investments of $50,000. The most valuable aspect of cash flow statements are that they can realistically tell you if you need more financing to keep from running out of cash.

Financial Strategies

had no cash flow left. It couldn't wait for a more opportune moment because its very existence was in danger.

Consider All Cash Flow Scenarios When Choosing a Security

To choose the right security for your company, examine various hypothetical cash-flow scenarios.

The methodology used to determine the best financing alternative begins with the development of financial forecasts based on different scenarios. All significant variables should be incorporated, including company-specific factors like sales projections and outside factors like interest rates.

MCI did not begin to use financial models until 1978, fifteen years after the company was founded. In retrospect, senior managers thought that financial models would have benefited their financing decisions and that taking the time to create them would have been worthwhile. That is because the large number of variables makes it difficult to make sound decisions without some sort of structured tool such as a set of pro forma financial statements. (A pro forma financial statement, such as a balance sheet or income statement, is a presentation of data using hypothetical numbers.) A pro forma balance sheet, for example, might reflect a potential debt offering, with the numbers adjusted accordingly. Managers use pro forma statements to help

Action Steps

1. Examine the health of your company's cash flow with a pro forma financial statement.

2. Forecast your company's liquidity over time using one-, three-, five-, and ten-year horizons. Use objective numerical measurements like the quick ratio (assets excluding inventory divided by liabilities).

3. If your forecasted cash-flow flexibility is unsatisfactory, write a detailed plan to correct the situation through cost-cutting, downsizing, and boosting income.

them understand the real impact of such a step on a company's overall financial situation.

Whatever the method used to examine cash flow, it cannot be emphasized enough that cash flow must be taken into account both with financial offerings and their effects on cash flow and in the simple day-to-day running of the business. Cash flow must be healthy and unencumbered. Your company's life depends on it.

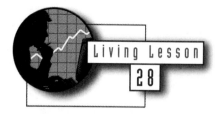

Living Lesson 28

Maximize Your Opportunities For Capital

In addition to timing your offering, it is essential that you capitalize on market demand and optimism when it is most favorable. Michael Milken always encouraged managers to raise as much money as possible when given the opportunity.

"One of my basic beliefs is that the best time for a company to raise capital is when it doesn't need it and the market is the most receptive," Milken explained. "If the economics are favorable and the time is right, a company should raise as much money as it can."

MCI followed this tenet many times, and in particular when it decided on its billion-dollar bond-warrant offering in 1983. Following the record-setting sale, some observers publicly took MCI to task for being "greedy." But as history shows, MCI was just

Action Steps

1. Where applicable, examine your most recent financing experiences. Depending on the size of your company, this could mean considering your last bank loan infusion or your last public offering. How did you arrive at the total amount? What period of time were you hoping to cover?

2. Did you receive as much money as possible? If not, why? (There may have been good reasons, of course, like low market demand or valuations, or bank loan limits, but absolute cash-flow needs should always take first priority. Remember, running out of cash is the only mistake from which companies cannot recover.)

3. Examine your company's current standing with various financing sources. Is this a timely point at which to obtain more financing even if you don't need it in the short term? For example, if your business is in an up cycle, this may be the right time to raise the credit limit on your company bank account or increase the amount of your loan.

being smart. The future is unknowable—regulations can change, the market can shift—so it is essential to finance ahead of your requirements by taking advantage of favorable market conditions when they exist.

For a company like MCI, engaged in an extremely capital-intensive business and devouring $50 million a month in the early 1980s, choosing to follow this Milken principle was a particularly astute management decision. Bill Conway, who eventually became MCI's chief financial officer, put it succinctly: "Get it while you can, and get extra."

Be Creative in Your Financing

The variety of instruments available to companies today can be overwhelming—but the overview can represent great opportunity if approached openly. MCI, with the help of its bankers, has never been afraid to be creative in the financing tools it uses to address its unique needs, whether with convertible securities, preferred stock, or more complicated hybrids. Here is a review of the reasoning behind two of MCI's more interesting financing choices.

Preferred Stock

MCI used preferred stock for its financing a number of times during its earlier growth years. Like common stock, preferred stock provided MCI with equity, but its method of payment was bond-like since preferred stockholders receive dividends in fixed amounts over time. The difference is that bond interest payments are tax-deductible for the issuing corporation; dividend payments are not. For this reason preferred stock is not a particularly popular method of financing except for companies that don't yet have net income and thus are non-tax-paying entities. Preferred stock can also be appropriate for companies that think their stock is ready to take flight, as was the case with MCI in 1980.

MCI wanted to sell equity to pay off its senior debt to the banks, and did not want to take on more public debt. It could have issued common stock, but the company was optimistic about its future and did not want to sell equity at the then-current share price. The time seemed right for a convertible preferred offering. The winds of Wall Street had started to shift, and new companies and new industries were beginning to have broad investment appeal. But most importantly, MCI had won its antitrust suit against AT&T.

By issuing convertible preferred rather than straight preferred, MCI retained the option of selling equity at a price higher than that at which it was then trading. As it turned out, MCI's stock did take off and within thirteen months MCI had converted almost all of its preferred stock into common stock.

Drawbacks

Although convertible preferred stock offerings were successful for MCI because it was poised for growth, they do have their down side. If the common share price does not increase, and the preferred is never converted into common stock, the fixed dividend payments can strain a company's cash flow, which is not a good situation for a business that has yet to turn a profit. However, a company does have the option of buying back the securities in the open market or exchanging them for another type of security. Moreover, failure to pay preferred stock dividends will not throw a company into bankruptcy or reorganization proceedings.

Bond-Warrant Unit

In 1983 MCI had a considerable equity position already, so obtaining financing through a straight stock offering was out of the question. "Dilution of the company was difficult because we had already diluted our company quite heavily," related CEO Bert Roberts. "We needed money, and we needed it to fit within a framework and a capital structure that made sense at that time." A variety of debt options were available, and Drexel weighed the benefits of all of them.

High-yield bonds were an option, but interest rates were at historic highs and MCI was reluctant to issue long-term debt at 14 percent. In addition, straight bonds would increase the company's debt-to-capitalization ratio to 70 percent, which would cause concern in the investment community regardless of the company's market value. Finally, straight bonds would not take advantage of MCI's robust stock performance.

A convertible bond would allow MCI to get a lower interest rate and a longer maturity than would straight debt. Paying 10 percent interest instead of 14 percent would reduce annual interest payments by about $40 million. Also, MCI had been successful with convertibles before. The company feared, however, that it may have been too successful. Just four months earlier MCI had completed a $400 million convertible bond offering and now was concerned about a potential lack of enthusiasm for a new offering of this type. In addition, $1 billion worth of new convertibles might drive down the prices of both outstanding convertibles and the company's stock price. That outcome would make for bad blood with investors.

Milken wanted an instrument that incorporated the best features of the other options while providing a safety net for MCI if the stock market crashed. Additionally, he wanted something that gave investors a measure of comfort as well as a high yield. Milken's recommendation: a bond-warrant unit.

Among its key advantages was its appeal to a wide range of investors, which broadened its marketability. The warrants appealed to equity buyers and the bond portion appealed to straight debt buyers. Furthermore, the unit as a whole was a synthetic convertible, making it attractive to convertible-security buyers. (It is called a synthetic convertible because, although the unit is not a standard convertible per se, the warrant portion can be converted.)

Another major benefit was that only $800 million of the offering amount was registered on the balance sheet as debt. The warrants—valued at $200 million—were booked as permanent equity. In comparison, a convertible debenture is booked as debt only.

Furthermore, issuing warrants didn't dilute MCI's equity nearly as much as a straight stock issue would have. And although the equity feature comprised just 20 percent of the unit, it still enabled MCI to take advantage of its stock performance because the warrants were exercisable at a 31 percent premium above the then-current trading price. In addition, the warrants provided some protection to investors in the event a market collapse made them unexercisable: Each one hundred warrants could be converted into one share of common stock. This feature even allowed MCI to declare the expiration as nontaxable, which provided the company with substantial savings.

Finally, the bond-warrant unit also carried a lower interest rate—9 1/2 percent—than other forms of debt financing. The next best thing would have been a convertible at 10 percent. And over ten years, those additional basis points would have cost MCI $50 million extra in interest payments.

Drawbacks

As with any type of financing option, there were drawbacks, of course. The bond-warrant unit didn't take full advantage of MCI's strong stock performance, and although the interest rate was relatively low, the bond yield was high at 13 1/4 percent. In addition, if all the warrants were exercised, MCI's equity would be diluted by about 15 percent. But Milken and MCI settled on the combination unit because it offered both the issuer and the investors the greatest benefits with a fairly limited down side.

Action Steps

1. Before considering a public offering, make sure you understand the basics of the various available financing instruments. Where applicable, have your investment banker summarize the pros and cons of the different instruments, taking into account your particular company's situation. Are they considering all of the options? Have they found the right fit for your company?

2. Pay attention to the financings of other companies in recent history, particularly those you believe to be analagous to your company. Ask yourself why they choose one option over another and how that choice would have worked for your company.

3. For smaller companies, list the financing alternatives which may be available to you outside of your standard sources. Consider all possibilities including partnerships with larger companies and grants from government programs. Could any of these options be more advantageous than your current sources?

Let the Market Value Your Company

Living Lesson 30

Although the investment community still takes a book value seriously, market capitalization is considered to be a more accurate reflection of a company's worth. (Market capitalization is sometimes referred to as "market value" although financial textbooks define them differently.) But what's the difference?

Book value is an accounting record of past equity investments in the company. The number basically reflects the shareholder equity at the price it was sold plus any profits that your company may have gathered in retained earnings. Market capitalization is the total dollar value of all of a company's common stock based on its current market price.

As you can see, the problem with book value is that it is based on a company's past. Market capitalization reflects how investors view a company's future. At the end of 1996, for example, MCI had a book value of $10.6 billion as reflected on its

> When you sell your house, you don't base the asking price on what you paid but on what the market is willing to pay.

Action Steps

(1) Calculate the book value of your company and, if possible, your market value. If you do not have publicly traded shares, approximate your company's market value using other benchmarks for companies of your size and in your industry. These may include standard ratios such as price-to-sales or price-to-earnings measurements.

(2) How do your book and market values differ? If possible, reevaluate your estimated market value periodically and incorporate it into your long-term strategy, including your exit strategy options.

(3) Where beneficial, incorporate your company's market value into your presentations to investors and customers. (Note: In this circumstance, you should ensure that your calculation is accurate and documented. Any estimating variables should lean toward the conservative side.)

balance sheet in shareholder's equity. By contrast, its market capitalization (using only its common stock) was $19.4 billion (share price of $32 11/16 x 593 million shares outstanding).

When you sell your house, you don't base the asking price on what you paid, but rather on the property's appreciated (or depreciated) value as determined by the market. Similarly, looking at what you paid won't show how leveraged you really are. For example, holding a $180,000 mortgage on a home for which you paid $200,000 would seem to indicate that you are highly leveraged. However, if your home has appreciated to $300,000, simply using the mortgage-to-purchase-price ratio will provide an inaccurate picture of your current debt-to-asset ratio.

In MCI's case the success of its 1983 offering clearly showed that the market did not think the company was overleveraged. MCI's book value equity was about $760 million and its debt $940 million, meaning that about half of the company's capitalization was comprised of debt. Using MCI's market value of about $3.3 billion, however, only 20 percent of its total capitalization was made up of debt. If the investment community had looked to MCI's book value only, the majority of MCI's financing events would have been unsuccessful.

The market is even less concerned with book value now than it was in 1983. Nothing illustrates this more strikingly than the white lightning surge of high-tech IPOs. Netscape Communications Corporation, the California-based software maker, was just fifteen months old and losing money, but it carried a market value of $463 million when it went public in August 1995. The IPO was priced at $28 a share; by December, the share price was up to $171.

Financial Strategies

Know Your Market for Money

Living Lesson 31

Corporations, money managers, individuals, and major financial institutions all buy financial assets such as bonds, stocks, and derivatives. It is the major institutions, however, including commercial banks, mutual funds, insurance companies, and retirement funds, that purchase the overwhelming share of financial assets.

Understand the Buyers

It is important to get to know the needs and financial goals of these buyers to make successful offerings. Each institution's needs and goals will be governed by a number of different factors including tax considerations, regulatory restrictions, maturity needs, and differing investment objectives. Once you understand the most important factors, you will be able to choose an instrument that the market is ready to buy.

The list of institutional buyers is fairly evenly distributed, with banks, insurance companies, mutual funds, and pension funds equally important as potential buyers. Each of these groups will have different restrictions and needs and should be addressed accordingly when marketing a deal.

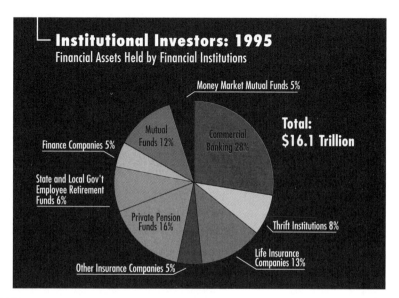

Institutional Investors: 1995
Financial Assets Held by Financial Institutions

- Money Market Mutual Funds 5%
- Mutual Funds 12%
- Commercial Banking 28%
- Finance Companies 5%
- State and Local Gov't Employee Retirement Funds 6%
- Private Pension Funds 16%
- Thrift Institutions 8%
- Life Insurance Companies 13%
- Other Insurance Companies 5%

Total: $16.1 Trillion

MCI: Failure Is Not an Option

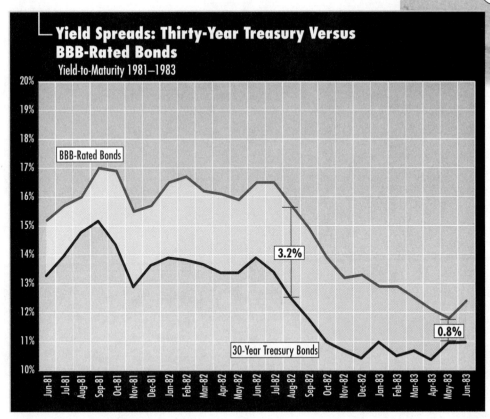

Watch the Yield Spreads

The spread is the difference in the return generated by different kinds of securities, typically issues of differing credit quality, such as the yield on long-term Treasury securities and that on corporate securities. If you want to sell debt, yield spreads must be carefully studied. Because yields are inversely related to bond prices, the spread is simply a measure of the relative prices of the securities.

If a ten-year government Treasury bond is yielding 7 percent to investors, for example, and a ten-year corporate bond is yielding 8 percent, then the spread is 1 percent or 100 basis points. When corporate bonds are seen favorably, the spread narrows, and when the risk is perceived as greater, the spread widens. Periods of narrow spreads, on a relative basis, are the most advantageous time for corporations to issue bonds.

Before MCI's July 1983 financing, yield spreads had narrowed substantially, from a 3.2 percent differential (320 basis points) in August 1982 to 0.8 percent (80 basis points) just before the offering. The narrower spread made a bond offering a more attractive financing option.

Financial Strategies

High-yield securities pay a higher rate of return to compensate for a greater risk of default. A narrow spread means that a high-yield security is viewed as not much more risky than a Treasury bond. Because the risk of default is perceived as smaller, an issuer does not have to offer as high an interest rate to attract investors.

Factor in Inflation

The rate of inflation is another important factor to consider before putting together your offering. At the time of MCI's 1983 bond-warrant issue, the U.S. economy was just emerging from an era of double-digit inflation during which bondholders had suffered significant capital losses. Even though the Federal Reserve had largely contained inflation by this time, investors were wary and still demanded inflationary premiums in bond yields. Since

Action Steps

1. Create a simple report on market conditions that can be updated periodically. It should include an update on general interest rates and spreads between various types of bonds (most importantly Treasury and corporate bonds). If possible, incorporate a simple chart that will give you an update at a glance of overall interest rate trends.

2. Arrange to receive a periodic report on general economic conditions through the research departments of your investment banking contacts, by subscribing to a general newsletter, or even by focusing on a recurring quick-summary feature in your favorite business periodical.

3. If you have already established a relationship with investment bankers, ask them for a summarized report on the demographics of likely investors in your company. The report should include factors like:

   - investor benchmarks (i.e., a required return on assets or sales growth)

   - investor restrictions (i.e., minimum bond rating or company size)

   - and general areas of interest (i.e., are there investment funds which focus on your industry?)

the prices of long-term bonds are more sensitive to inflation, their yields at that time tended to be especially high.

Unless they have a call provision, long-term bonds carrying high rates of interest can be a problem for an issuer. If interest rates decline, the issuer is still locked into paying out high rates when lower ones are available using shorter-term financing. Because many issuers were caught in just such a situation in the 1970s, shorter maturity lengths became the norm in the following decade. Corporate debt issues in the 1980s rarely carried maturities in excess of ten years. (Convertible securities, which offer corporations an early out, might extend to fifteen to twenty-five years.)

Inflation has been less of a concern in the 1990s, but every company should be mindful of general interest rate trends in planning its financing future.

Market Your Offering Actively

Living Lesson 32

All but the most dedicated number-crunchers will tell you that decimal points and percentages alone can be tedious. That is why it's important for management—the chief executive, the chief financial officer, the chief operating officer, and the treasurer—to market any new issue by personally visiting institutional investors.

MCI's 1983 offering involved more than two hundred institutional buyers, and included some of the biggest names in the business. Their enthusiasm for the issue was no accident, nor was it solely the result of MCI's numbers and projections.

> Investors like to look someone in the eye before they place their confidence—and their money—in a business.

Connect With Potential Investors Personally

Certainly MCI management knows how to court potential investors as well as anyone. Its executives participated in every Drexel High-Yield and Convertible Bond Conference after Drexel became its investment banker. At its peak the conference attracted more than three thousand institutional investors from

Action Steps

1. With the help of your investment bankers, arrange an informal outing—a company visit, or recreational event—at which your company can mix with potential investors.

2. For smaller companies, concentrate on maintaining close contact with your investors or financing sources. Arrange company visits or periodic dinners to keep them personally interested in your company's progress.

3. Consider inviting investors to important company events or milestones—the unveiling of a new product, for example, or celebration of surpassing sales objectives.

around the world. McGowan was so keen on connecting with these people that he even attended the 1988 conference, less than a year after his heart transplant. Although McGowan's diligence was exceptional, persistent visibility is crucial in winning the confidence of institutions like these.

Drexel often set up breakfast or lunch presentations where MCI could mix with potential investors. Whenever possible, the firm bowed out of the meetings and let MCI do the talking. These visits gave management an opportunity to discuss its vision and strategies without having a go-between interpret, and possibly misrepresent, the company's perspective on the future.

Like everyone else, institutional investors appreciate the time and effort involved when management comes to make a presentation. They like to look someone in the eye and hear a voice before they place their confidence—and their money—in a business. Beyond the personal touch that these meetings provide, they're good opportunities for management to make contacts, regardless of whether they sell the issue. Meetings open doors, and investors may later remember you when it comes to future trades.

Good marketing and investor relations are now more important than ever. Today, the environment of disclosure necessitates keeping the investor informed. Management's ability to develop a close, open relationship with institutional investors can provide a foundation for a long-term working partnership that will reward both the investor and the issuer.

Living Lesson 33

Solicit Help From Outside Sources

Savvy CFOs talk to a variety of financial experts before committing to a course of action. Besides dispensing the services they advertise, professional firms in accounting, banking, and law can provide advice and direction on many issues with which your company may be unfamiliar.

Accounting firms, for example, can do more than help keep books straight. They can also assist in determining the value of a company's assets, which is a highly specialized task.

Smaller companies unable to afford big staffs aren't the only ones that should consider such services. Big companies may find them valuable, too. At one time MCI narrowmindedly solicited only services that fell within the strict parameters of a company's business. "Today," said Roberts, "we look at these firms in terms of strategic expertise."

MCI has turned to major investment banks to help develop its company strategies, not just to bring money into the company. "We've also looked to investment banks—or any of our

Action Steps

1. Develop a relationship with at least one trustworthy contact at all companies that service your company, including your law firm, accountants, bank or banks, advertising agency, and vendors.

2. Solicit the advice of these contacts on a regular basis as to how you might improve your company or further its goals. Open the door for any ideas they may have.

3. Chart these ideas in a grid organized by topic (such as new issues, expansion, acquisitions, and so forth) and look for duplication of ideas by people from different disciplines. These ideas will more than likely be worthwhile to pursue.

other professional services firms—to provide independent advice on major transactions, such as a joint venture or acquisition," Roberts said.

Another advantage is that these firms have an extensive reach into a number of industries because of the deals they have structured. They can be experts in locating contacts for you.

Because of the complex nature of many of these deals, services among the major professional firms have blurred in recent years. For example, law firms might offer services that overlap with those of accounting firms.

MCI's 1993 alliance with British Telecom was part of a deal in which MCI received $4.3 billion for 20 percent of the company. "It was a complex transaction," Roberts explained, "involving not only all of the usual terms that go with a transaction of that nature, but also how we would deal with all of the regulatory agencies around the world." Simpson Thatcher, the company's law firm, worked out both the business and the legal issues involved in the deal.

The bottom line, of course, is that a young company should be open to soliciting and accepting help from all quarters and will often find it in unexpected places.

Index

A

accounting firms, 338–339
Advertising Age, 130, 131
AdWeek, 152
Aeroquip Corporation, 58
Airfone, Inc., 50, 51
Airsignal International, 127–130, 238
Allen & Company, 68, 319–320
Allen, Robert, 163, 187–188
alliances, 262–264
Ally & Gargano, 70, 72, 74, 147, 282, 283, 296
Amdahl Corporation, 53
America Online, 317
American Academy of Arts and Sciences, 180
American Association for the Advancement of Science, 180
American Bell Telephone, 29, 221
American Express, 75
American Marconi, 217
American Marketing Association Award, 285
American Paging, 127
American Sky Broadcasting (ASkyB), 193, 246
American Speaking Telegraph Company, 29
American Telephone & Telegraph (AT&T), xii, 29, 33, 35, 37, 38, 48, 49, 71, 73, 80, 81, 98, 102, 104, *125*, 130, 137, 159, *161*, 163, 167, 174, 176, 178–179, 183, *186*, 186–188, 194, *200*, 204, 206, 216–218, 221, 222, 239, *241*, 243, 275, 282, 292, 293, 316
    divestiture, xvi, 79, 113–117, 121, 132, 135, 149, 163, 186, 196, 223, 236, 237
    interconnection access for MCI and, 28–30, 32, 39, 41–43, 55, 60–61, 63, 110, 117, 135, 178–179, 184, 192, 227, 231, 232, 268
    lawsuits. See antitrust suits vs. MCI, 12, 13, 14–15, 18, 19, 22, 25–27, 54, 60–63, 94–95, 135, *142*, 147–150, 152–153, 156–157, 171–172, 173, 207, 225, 227, 228, 229, 231, 232, 233, 258, 265, 266–267, 279–281, 283, 286–287, 294

Ameritech, 118, 218
Amtrak, 99
analysis of financing factors, 303–307, 318
"Anna Campaign," 168, 169
antitrust suits, 41, 43, 54, 60–61, 69, 77–78, 79, 83, 113, 120–121, 148, 184, 216, 217, 222, 227, 232, 233, 238, 268–269, 306, 326
ARPAnet, 181
Asia, 158, 263
Association for Computing (ACM), 180
ATM generation, 144
Australia, 102
Avantel, 47, 143, 159, 196
Avis, 74

B

Bader, Michael, 19, 28, 42
Bank of America, 160
banks and bankers, 34–35, 43, 55, 56–60, 63, 65, 67, 68, 84, 88, 91, 104, 160, 226, 230, 232, 308, 312, 313, 314–315, 318, 326, *332*, 334, 336, 338
Barrett, Leonard, 17
Bear, Stearns & Company, 185
Belgium, 102
Belize Telecommunications, 183
Bell Atlantic, 118, 193, 210, 218
Bell Laboratories, 98, 113, 187, 216, 221, 258
Bell South, 118, 218
Bell Telephone Company, 29, 215–216

Bell, Alexander Graham, 29, 215, 219, 221
Bendix Corporation, 196
Berger, Ron, 72
billing services, 44–49, 137, 139–140, 202, 203, 239, 246, 275–277
Billion Dollar Day, *171*
Blyth & Company, Inc., 35
bonds, 66, 85–93, 95, *96*, 103–107, 233, 234, 235, 236, 238, 263, 303, 306, 308–309, 310, 312, 327, 328, 329, 331, 333–335
bond-warrant financing, 103–107, 110–111, 307, 310, 324–325, 327–329, 334–335
book value, 330–331
Bouygues, *201*
brand-name recognition, 156, 169, 273–274
Brier, Danny, 182
Briggs, Fred M., 194, *194*
British Post Office, 204
British Telecommunications plc (BT), xiv, 26, 46, 83, 102–103, 131, 159, 176–177, 178, 185, *200*, 204, 210, 218, 243, 263, 339
broadcasting, 100, 191, 217
Bruegge, Carl Vorder, 250, 298
Buchan, Alexander, *27, 31*
Budget Rent-A-Car Corporation, 185
Burnham, I. W. (Tubby), 89
Burns, George, 154
business representation, 173–174, 288
business service/market, xv, 153, 167, 169, 172–174, *182*, 183, 193, 198, 202, 203, 211, 242, 244, 263, 287, 288, 294–295, 296–297, 298–300
Business Week, 169, 173, 195

C

Cable and Wireless plc, 159, *201*
cable companies, 218, 219, 269
cable service, 79, 191, 204, 307
Cablevision Systems, 90
Caliber Learning Network, 198
California, xviii
calling card, 101, 195
Canada, 83, 101–102, 143, 178, 204, 236
capital and capital markets, xvii, xviii, 10–11, 14, 20–22, 34–35, 55–56, 58, 65–68, 80, 83–97, 103–111, 126, 129, 133–134, 177, 207, 226, 229, 232, 233, 234, 235, 236, 238, 262, 263, 264, 268, 324–325
  general financial strategies and, 303, 305, 306–313, 317–339
capital structure, 85, *86*, 91–92, 311–314, 317, 318, 327
Carey, Wally, 72
Carlyle Group, 84
Carter administration, 196, 270
Carter, Tom, 15, *223*
cash flow, 45, 303, 307, 312, 317, 318, 321–323, 324, 327
CBS, 147
CD-ROM, 197
Cegetel, *200,* 204
cellular service, xix, 127–129, 183, 195, 204, 211, 217, 219, 238, 245, 246, 307
Centel, 218
Central America, 183
Cerf, Vinton G., 179–181, *180*
Charge and Save telephones, 101
Charms, 119
China, xv, 205
Cigna, 107
citizen-band radio, 218
City Investing, *320*
Clear, *200*

Clinton, Pres. Bill, 47, 180, 191, 219, 223, *245*, 268
CML Satellite Corporation, 30–31, 36, 44, 125, 130, 228, 262
collect calls, 137, 169, 174–176, 239, 243, 279, 291–293. *See also* 1-800-COLLECT
Collins Radio Company, 33, 38, 43
Columbia Savings & Loan, 107
COMAC, 9–10
Comcast, 183
common stock, 55–56, 65–66, 107, *108–109,* 307–308, 309, 311, 313, 320, 326, 327, 328, 330–331
Communications Act of 1934, 217, 222
Communications Consultants, Inc., 17
Communications Satellite Corporation (COMSAT), 30, 36, 228
competition, xvi–xvii, xix, 75, 79, 102–103, 129, 132, 135, 150, 158, 162–163, 148, 188, 199, 204, 206, 211, 223, 225, 228, 252, 256–257, 258, 263, 273, 279–281, 299, 304, 306
computers, xix, 51, 113, 177, 195, 197, 216, 263, 297
Concert alliance, 83, 131, 176–177, 184, 204
Concert Communications Company, 143, 177
Concert InternetPlus, 26, 246
conference calls, 160
conferencing division, 183
Congress, 42–43, 227, 269, 270
consumer division, 151, 154, *See* MCI Consumer Markets
content, 182, 193, 205
Continental Telephone, 38, 89
convergence, 46, 77, 177–178, 193–197, 207, 244
convertible bonds, 88–89, 91–92, 95, *96, 105,* 235, 308–309, 310, 328, 335

convertible debentures, 307
convertible preferred stock, 65–68, 84, 91, *108,* 232, 233, 327
convertible subordinated debentures, 91–93, *109*
Conway, William E. (Bill), 83, 84, *84,* 89, 91, 93, 107–109, 111, 123, 249–250, 325
Coopers & Lybrand, 58
Corestates Financial Corporation, 185
Corning Glass, 97
Corning Incorporated, 198
corporate culture, xiv–xv, 23–24, 25, 31, 33, 39, 47, 62, 100–101, 119, 120, 141–144, 157, 181, 197, 207–208, 210, 212, 247, 255, 261, 283
Corporate Office, *123*
Corporation for National Research Initiatives (CNRI) , 180
Corporation for Public Broadcasting (CPB), 25
courier service, 297
Cowett, Ed, 9, 10, 13
Cox Communications, 183, 219
Cox, Kenneth A., 42, 269, 270
credit card calls, 160
credit card charge processing, 205
CRICO Communications, 32
Crown Center Redevelopment Corporation, 58
Crystal Oil, *320*
CSX Corporation, 99
customers, 258–259, 261, 294, 296–297, 299
customer service, xiii, 73–74, 83, 137, 140, 195, 200, 202, 203, 205, 290, 296
customer-service call centers, 205
CyberEd, 197–198
CyberRig, 197–198

D

Dallas One, 50, 229
Darome Teleconferencing, 183
data transmission, xv, 44, 83, 100, 113, 157, 158, 159, 174, 177, 193, 201, 203, 240, 263
Datran, 217
Deatherage, William, 185
debt financing. See bonds
debt-to-asset ratio, 331
debt-to-equity ratio, 317
deButts, John, 30, 130, 239
decentralization, 46, 117–120, 131
"Defining the Future" campaign, 169, 170
delegation, 253–254
Department of Commerce (state), 266
Department of Justice, 18, 42, 79, 113, 217, 269
deregulation, xvi, 187, 191–192. See also regulatory issues; Telecommunications Act of 1996
Deutsche Telekom, 159, 176, 177, 200, 201
digital television, 100
digital transmission, 51, 128, 167, 168, 170, 178, 203, 242
direct-broadcast satellite (DBS) services, 193, 199
disclosure, 318, 337
discounting, 279–280, 286
diversification, 181–186
diversity, xiv–xv, 260–261, 277–278
divestiture. See under American Telephone & Telegraph (AT&T)
dividends, 65, 307–308, 326, 327
Donoghue, John, 120, 175, 277–278
Drexel Burnham Lambert, 58, 87, 88–91, 95, 103–107, 122, 126, 138–139, 234, 327, 336–337
Drexel High-Yield and Convertible Bond Conference, 126, 138–139, 336–337

Dunkin Donuts, 72
Dunlap, Angela, 101, 120, 129–130, 131, *131*, 142, 155, 156, 168, 186, 284

E

Ebbers, Bernard, xiv, 212
economic conditions, 90, 306, 334
Edison, Thomas A., 216, 219, *221*
education, technology use in, 47, 197–198
Eidenberg, Eugene, 270–271
800 service. See toll-free 800 service
e-mail, xix, 139, 158, 168, 170, 178, 195, 204, 246, 297
emotions, 299
employees, xvi, 52–54, 56, 62, *65*, 140–141, 149, 151–152, 157, *164–165*, 173, 248, 249–250, 253–257, 260–261, 270, 280
English, Wayne, 57–59, *58*, 65–68, 83, 85, 89, 91, 93, 103, 104–105, 110, 123, *230*, 250
entrepreneurism, 50, 54, 88
Environmental Protection Agency, 266
equal access, 63, 113–115, 121, 130, 135–136, 138, 196, 206, 237, 252, 277
Equal Access Consent Decree, 237
equity, 67, 83–84, 85, 88, 95, *96*, 103, 105, 110, 303, 307–308, 309, 310, 312, 313, 317, 320, 321, 326, 327, 328, 329, 330, 331
Europe, xv, xviii, 13, 102, 103, 158, 159, *201*, 204
European Union, 103, *200*
Exchange Network Facilities for Interstate Access (ENFIA), 61–63, 232
Execunet, 43–50, 54–56, 59–60, 61, 63, 64, 78, 82, 94, 97, 152, 225, 228, 229, 230, 231, 275–276, 286, 289, 294–295, 318
exit strategy, 330

Exxon, 122

F

fax service, 44, 83, 139–140, 141, 178, 197, 203, 240, 257, 263
Federal Aviation Administration (FAA), 170, 194, 242
Federal Communications Commission (FCC), 10, 21, 22, 30, 32, 39, 41, 48, 51, 54, 56, 60, 61, 62, 79, 81, 110, 115, 127, 135, 150, 171, 184, 191–193, 211–212, 216, 217, 218, 222–225, 230, 243, 265–267
  first MCI application before, 12, 13, 14–15, 19–20, 25–28, 224, 225
Federal Energy Regulatory Commission, 266
Federal Express, 72, 74
Federal Reserve, 266, 334
fiber optic network, xv, 83, 94, 97–99, 100, 133–135, 137, 147, 158, 159, 160, 170, 178, 179, 182, 201, 235, 239, 240, 241, 244, 258
Fidelity Management, 107
financial models, 322–323
Financial Women's Association, 130, 131
Financial World, 131
financing strategies, 303–339. See also capital and capital markets
First National Bank of Chicago, 34, 83, 84, 226, 318
Fleet Call, 219
flexibility, 258–259
focus groups, 74, 296–297
Food and Drug Administration, 266
Forbes, 182, 197
Ford Motor Company, 57, 58
Fortune, 169, 173
4K Plus, 44, 48, 64, 227
Frankfurt, xv
Fox network, 182
France, *200*, *201*

France Telecom, *200, 201,* 204
franchise model, 21–22, 266
French Telecom, 177
Friends & Family Connections, 131, 150–157, 169, 185, 202, 203, 205, 225, 237, 241, 254, 259, 274, 277–278, 280, 283, 287, 290, 291, 292
Frohlinger's Marketing Report, 143
FTD (Florists Transworld Delivery), 50

G

Gabbard, O. Gene, 160
Gabor, Zsa Zsa, 154
Gallagher, Edward, *81*
Garlinghouse, F. Mark, 28–29
Garthe, Kenneth, 17
Gates, Bill, 219
Geneen, Harold, 321
General Electric, 17, 83, 157, 240
General Electric Credit Corporation, 107
General Motors, 6, 75, 294–295
Geostar, 219
Geriatric Services, 10
Germany, xv, 6, 159, *200*
Gin, Sue Ling, 5, 131, 133
Global Communications Service, 167
Global Information Solutions, 187
Global One Alliance, *200, 201*
Goeken Group, 51
Goeken, John D. (Jack), 12, *13,* 17–22, *18,* 26, 27, *28,* 28–29, 44, 50–52, *51, 224,* 228, 260, 265
Goldberg, Whoopi, 155
Golden Nugget, 104
"Good Morning America," 168
Gore, Al, *191,* 245
government. *See* regulatory issues
Graham-Willis Act, 222
Great Britain, 137

Greene, Judge Harold, 79
Grupo Financiero Banamex-Accival (Banamex), 83, 159, 178
GTE, 51, 62, 74, 210
GTE AirFone, 51

H

Haley, Andrew G., 19
Hallmark Cards., Inc., 58, 59
Haloid Company, 82
Harris, Laurence E. (Larry), 28–29, 32, *32,* 41, 42, 78, 130, 249, 269–270
Harris, Louis, 291
Harvard Business School, 6, 6–7
Hermitage Holding Company, 196
Hertz, 74
High Performance Communications Office, 180
high-yield (junk) bonds, 85–88, *87,* 90, 95, 103, 107, 306, 327
Hilton International Corporation, 58
hiring strategies, 249–250, 260–261
Hirsh, Irwin, 226
home security service, 195
home shopping, 251
Home Shopping Network, 137
Hong Kong Telecom, *201*
Hughes Aircraft, 38, 100
Hughes Communications, 235

I

IBM, 44, 52, 53, 122–127, 225, 228, 229, 238, 259, 262–263, 275–276, 298, 300
ICF Kaiser International, 196
IDC, *200, 201*
Indiana-Ohio regional carrier, 24
inflation, 334–335
InFlight Phone Corp., 50, 51, 194
INFONET, 159
Information Age, 77, 168, 258

information technology, 90, 179–181
initial public offering (IPO), 34, 35–37, *108,* 226, 305, 313, 317–318, 330
Institute of Electrical and Electronic Engineers (IEEE), 180
institutional investors, 93, 104–106, 107, 332, 336–337
insurance companies, 88, 308, *332*
integrated services, 195, 205, 206, 246, 279
Intel, 26, 47
interconnection access, 28–30, 32, 39, 41–43, 55, 60–61, 63, 110, 117, 135, 178–179, 184, 192, 227, 231, 237, 268. *See also* equal access
Interdata Communications (regional carrier), 24
interest rate, 67, 68, 84, 88, 238, 303, 306, 308, 309, 312, 327, 328, 329, 334, 335
Internal Revenue Service (IRS), 266, 308
international business services, xv, 176–178
international record carrier (IRC), 80–81
International Telecom Systems, 32
international telecommunications, 140, 157–159, 176–178, 199, 205
International Telephone & Telegraph, 62
international telephone service, xv, 73–74, 80, 101–103, 137, *200–201,* 219, 234, 239, 240, 263, 277–278
Internet, x, xii–xiii, xv, xvi, xviii, 26, 47, 143, 168, 178, 179–181, 193, 195, 197, 198, *201,* 206, 211, 219, 244, 245, 246, 305
Internet 2000, 245
InternetMCI, 179–181, 244
intranets, xvi, 26, 205, 206
investment opportunities, xvii
Iridium, 219
ITJ, *200*

ITT Corp., 113, 321
ITT Dialcom, 204

J

Japan, 11, *200, 201*
Jenner & Block, 12, 43, 184
Jim Walter, *320*
Johnson & Johnson, 262

K

Kagan Telecom Associates, 33
Kagan, Jeffrey, 33
Kantor, Nathan, *81*
KDD, *200*
Kemper, 107
Kentucky Central regional carrier, 24
King's College, 6
Kingsberg, Malcolm, 7–8
Kingsbury Commitment, 216, 217
Klausen, Will, 140

L

Lancaster, Burt, 75, *75*
languages, 277–278
large accounts, 298–300
law firms, 69, 338–339
layoffs, 56, 81–82, 130–131, 252
LDDS (Long-Distance Discount Service), 212
Leasco, 32, 46, 298
Leased Interfacility National Air Space Communications System (LINCS), 170, 194
Lee, H. Rex, 25
Lehman Brothers, 207
Leming, Thomas L. (Tom), 38, *38*, 97–98, 100, 101, 250
Lenkurt Electronics, 38
Leno, Jay, 155
Letterman, David, 175, 292
Liebhaber, Richard, *123*
LIN Broadcasting, 218

liquidity, 317
Litton Industries, 113
loans and lines of credit, 34–35, 43, 55, 56–60, 65, 67, 68, 84, 88, 91, 160, 226, 230, 232, 308, 312, 313, 314–315, 318
local service, x, xii, xvii, 26, 113, 143, 184, 185, 191, 192–193, 199, 206, 210, 211, 216, 244
Lockheed Aircraft Corporation, 30, 31, 225, 228, 251, 262
Lockheed Martin, 31
London, xv
Long Lines Division (AT&T), 63, 228
long-distance carriers, 114–117, 132, 178–179, 199, 237, 243
long-distance cost/price, xvi–xvii, 22, 44–45, 62, 63, 73, 75, 117, 135, 156, 169, *182*, 205, 216, 237, 267
long-distance market size, *138*
long-distance service, x, xii, xvi–xvii, 49, 79, 107, 113–117, 129, 147, 156, 169, 195, 198, 202–203, 206, 211, 229, 246, 263, 290, 294
Lucent Technologies, Inc., 187
Lybrand, Ross Bros. & Montgomery, 58

M

Magna-Theatre, 8
Maine, Douglas L., 124, 125, 177, 207, *231*
Mainline Electronics, Inc., 17
Malone, John, 269
management information system department, 154
management strategies and action steps, 248–271
Marconi, Guglielmo, 217, 218, *221*
market capitalization, 65, *92*, 330–331
market share, xii, 116, *138*, 149, 153, *154*, 156, 161, 164–165, 172, 176, *186*, 188, 195, *200, 201*, 206, 232, *241*, 279–280, 286, 290

marketing and advertising, xviii–xix, 49, 69–75, 83, 101, 102, 116, 127, 142, 147–157, 167–170, 171–176, 185, 197, 199, 202, 225, 233, 237, 241, 253, 261, 262
  strategies and action steps, 273–300
Martin Marietta, 31
Maybell, T. Mark, 199
McCaw Cellular Inc., 90, 128, 129, 183, 186, 204, 218, 223, 238, 307
McCaw, Craig, 128, 129, 219
McDonald's, 282
McGowan, Andrew, 3
McGowan, Bill, viii, ix, xiv, 1, 3–12, *13*, 16, *21*, 27, *31*, 40, *41*, 41–43, 44, 46, 48, 50, 52, 56, 57, 59, 64, 69, 74, 77, 80, 81, *81*, 83–84, 88, 89, 91, 92, 94, 95, 100, 101, 103, 104, 118, 121, 122, *123*, 125, 126, 127, 130–133, 138–139, 140, 141, 142, 146, 147, *146*, 149, 158, 160, 161–164, 166, 208, *224, 227, 229, 234, 236, 239*, 239, 241, 242, *243*, 249, 250, 251, 252, 253–257, 258, 260, 265–266, 267, 268, 269, 283, 285, 300, 302, 318, 321, 337
  early days with MCI, 11–15, 20–25, 26, 28, 30, 32
  management philosophies, 7, 10–11, 23, 39, 52
McGowan, Katherine, 3
McGowan, Monsignor Andrew Joseph, 4, 5, 9, 11, 13, 30
MCI Communications Corporation, 33, 84, 143, 196
"MCI Connections" campaign, 170
MCI Consumer Markets, 131
MCI Decision, 225
MCI Digital Information Services, 180
MCI Fax, 240
MCI International, 81, 83, 101, 123, 139, 158, 194, 240
MCI Mail, 139, 147, 180, 236, 297
MCI Mid-Atlantic Communications, 27

MCI One, 143, 195, 205, 246
MCI Paging, 185
MCI Preferred, 167
MCI Systemhouse, 26, 143
MCI Telecommunications Corporation, 32, 84, 131, 143, 237
MCI Metro, 244
MCI-WorldCom, xii–xv, xvii, *201*, 206, 210–212
McMaster, Jack, 293
McNamee, Louise, 72
media-week magazine, 143
Melman, Larry (Bud), 175, 292
Meltzer, Linda, 195
Mercury Communications, 103, *200*, *201*, 204
Merrill Lynch & Company, 199
Messner, Tom, 70, 72, 73, 74, 148, 168, 170, 283–284, 285, 286
Messner Vetere Berger McNamee Schmetterer Advertising Agency, 19, 72, 273, 283, 286
Metromedia, *161*, 218, *241*
Mexico, xviii–xix, 47, 83, 143, 159, 178, 241
mezzanine financing, 312
MGM/United Artists, 104
Michigan (regional carrier), 24
Microsoft, 26, 47, 193, 198, 202, 203, 219, 245
Microsoft Network, 202
Microwave Communications, Inc., 12–22, 217, 223, 224
Microwave Communications of America (MICOM), 22–33, 35, 224
microwave network, 11–12, 17–20, 25, 27, 28, 38–39, 63–64, 91, 93, 94, 97, 99, 130, 167, 170, 217, *231*, 256, 258
Mid-Atlantic (regional carrier), 24, 120, 150
Mid-Continent Communications (regional carrier), 24
Mid-South (regional carrier), 24
Milken Cube, 303, *303–306*

Milken Family Foundation, 198
Milken, Michael, 58, 88–97, *90*, 103–106, 107, 109, 111, 126, 128, 162, 234, 303, 324, 325, 328, 329
Mirage Resorts, 104
mission statement, 299
mistakes, 251–252
Mitel, 204
Mobil Chemical Company, 194
Mobile Communications, 218, *320*
mobile radio, 219, 223
mobile-telephone service, 218
Modified Final Judgment (MFJ), 79
money-transaction services, 181, 185
Moody's Investors Service, 95, 103, 207, 235
"More Changes Imminent," 31
Morgan, J. P., 6, 222
Morse, Samuel F. B., 215, *220*
motion pictures, 216
Motorola, 38, 219
Mountain Bell, 69, 73
multimedia applications, 178, 197–198, 246
Murdoch, Rupert, 181–182, 193, 244
mutual funds, *332*

N

Nacchio, Joe, 281, 288
NASDAQ, 67
National Accounts Management Association, 143
National Association of Regulatory and Utility Commissioners, 192
National Geographic, 99
National Rules, 184
National Telephone Company, 204
Nationwide Cellular Service, 183, 245
nationwide service, 37–39
NCR Corporation, 187

Net Day '96, 47
NET Vote '96, 47
Netscape Communications Corporation, *36*, 305, 317, 331
Network Systems division (AT&T), 186–187
networkMCI, 143, 177–178, 183, 244, 279
New England (regional carrier), 24
New York Daily News, 182
New York Times, 139
New York West (regional carrier), 24
Newcomer, Judge Clarence, 41
News Corporation, Ltd., 26, 90, 181–183, 193, 196, 244, 246
Newsweek, 293
Nextel Communications, 128, 219
900 service, 139, 240
Nippon Telephone and Telegraph, 176, *200*
North-Central States (regional carrier), 24
Northwestern Mutual, 107
NYNEX, 118, 210, 218

O

O' Neil, Gerard, 219
O' Neill, Jr., Thomas P. "Tip", 10, 270
Office of Corporations, 266
Office of Telecommunications (OFTEL; Great Britain), 204
Office of Uniform Commercial Code, 266
Oklahoma, 8
1-800-COLLECT, 169, 174–176, 243, 279, 291–293
1-800-MUSICNOW, 251
1-800-OPERATOR, 279
on-line access, 177–181, 191, 193, 197, 244. *See also* e-mail, Internet
on-line shopping service, 244, 251
operator services, 137–138, 239

Oppenheimer, 320
options, 310
Optus Communications, 201

P

Pacific Coast (regional carrier), 24
Pacific Mountain States (regional carrier), 24
Pacific Telecom, 320
Pacific Telesis, 118, 218, 219
Paging Network, Inc., 127, 183–185
paging service, 127–129, 178, 183–185, 195, 218, 238, 246
PaineWebber, 35
Pakistan, 201
Paktel, 201
Paquin, Anna, 168, 169
Paris, xv
pay-in-kind securities, 307
pension funds, 332
Pentagon, 137
People, 180
PepsiCo, 35
personal communications services (PCS), 219
Philips, Donald, 17
Philips, Nicholas, 17
Pony Express, 215, 219, 220
Popek, Beverly, 112, 120
postal, telegraph, and telephone ministries (PTTs), 102, 158
Powertron, 9
preferred stock, 65–68, 84–85, 91, 92, 108, 232, 233, 311, 313, 320, 326–327
Price, Timothy F. (Tim), 72, 73, 119, 120, 139, 141, 143, 143, 147, 150, 151, 153, 154, 157, 165, 168, 172–174, 175–176, 199, 203, 237, 250, 251–252, 257, 279, 280–281, 287, 288, 290
price-to-earnings ratio, xvii, 330
price-to-sales ratio, 330
private-line service, 44, 45, 56, 61, 78, 217, 227, 228, 230
procedures, 7, 11, 52
Procter & Gamble, 273, 274
product launch, 294–295
professional services firms, 338–339
pro forma financial statement, 322–323
"Proof Positive" campaign, 172–174, 237, 242, 288
PTAT-1, 159
publicity, 274
Pullman Company, 57
"Put It in Writing" campaign, 148–150, 286, 287

Q

quality of connection, 98, 135
quick ratio, 322
radio, 216, 217, 221, 222

R

Radio Corporation of America (RCA), 52, 53, 217, 229
railroad rights-of-way, 98–99, 133
Ransohoff Company, 196
Raytheon Company, 33–34, 226
RCA Global Communications, Inc. (Globcom), 83, 157–158, 240
Reach Out America, 147, 148
Reagan, Ronald, 306
regional Bell operating companies (RBOCs), 15, 18, 24, 28, 61–62, 79, 113, 115, 116, 117, 118, 121, 127, 178–179, 184, 185, 187, 192, 199, 204, 206, 210, 217, 219, 232, 236, 263
regional holding companies. *See* regional Bell operating companies
regulated monopoly, 29, 216, 222
regulatory issues, x, 10, 13, 14, 81, 83, 184, 191–193, 199, 216, 217, 219, 265–271, 304, 306, 318, 332. *See also* Federal Communications Commission (FCC), state regulatory commisions, Telecommunications Act of 1996
remote access, 44
residential service, xvi–xvii, 69–75, 80, 147–157, 170, 182, 198, 233, 276, 282, 286, 287, 296–297
restructuring, 46, 131, 237
retained earnings, 330
revenues, 44, 45, 48, 50, 54, 60, 64, 65, 67, 78, 95, 118, 125, 136, 142, 148, 152, 156, 164–165, 187, 198, 202, 205, 229, 293
Revlon, 35
Ritz, 151
Rivers, Joan, 283
RKO, 7
Roberts, Jr., Bert C., xiv, 1, 35, 44–49, 46, 53, 54, 63, 76, 101, 103, 107, 118, 121, 123, 126, 131–132, 139, 141, 150, 152, 153, 154, 155, 160, 161, 164–165, 178, 179, 182, 186, 195, 197–198, 211, 212 213, 226, 231, 242, 249, 255, 259, 261, 264, 327, 338–339
Rohm & Haas, Inc., 44
Rohr Industries, 320
rotary-dial telephone, 114
Rowny, Michael J., 196, 196

S

sales teams, 299
Salsbury, Michael H., 184, 184
Sarnoff, David, 217
Satellite Business Systems (SBS), 122–127, 130, 238, 262–263
satellite communications, 30, 99, 99–101, 100, 128, 193, 198, 219, 225, 235, 246, 258, 269–270
satellite television service, 100, 193, 199, 245
SBC Communications, 218, 219
Scheinman, Stan, 34, 35, 57, 226, 251, 252, 315, 318
Schmetterer, Bob, 19, 33, 72, 140, 156, 168, 169, 170, 173, 174, 175, 176, 202, 273, 274, 288, 296
screen-pops, xiii

Securities and Exchange Commission (SEC), 36, 104, 105, 266

securities offerings. *See* capital and capital markets; specific financial instruments

Securities Office, 266

senior cumulative convertible preferred stock, 67–68, *108*

senior debt, 88, 308, 311, 312, 326

senior secured debt, 85, 88, 311, 312

Shearson Loeb Rhoades, 68, 91, 233, 234

Shearson/American Express, 105

Shell Oil Company, 7

Sherman Antitrust Act, 43, 77, 227

SHL Systemhouse, 26

Silicon Valley, xviii

Singapore Telecom, *200*

"60 Minutes," 98

Skibo, Charles, 117

Skouras, George, 8

SkyMCI, 193, 246

SkyTel Corporation, 183–185

Sloan, Alfred, 6

social trends, 89, 305–306

software, xviii, 100, 193, 216, 244

solar-energy collectors, 216

SONET, 178

South America, 183

Southeast (regional carrier), *24*

Southern Pacific Communications, 62

Southwestern Bell, 218, 219

Spain, *201*

Specialized Common Carrier Decision, 225

Spectrum Analysis & Frequency Planning, Inc., 43

Spirit of MCI Award, 140

St. Louis–Texas (regional carrier), *24*

Standard & Poor's, 95, 103

state regulatory commissions, 184, 192–193

Stein Roe, 107

Stentor, 83, 143, 178, 196

STET, *201*

stock offerings, 34, 35–37, 55–56, 65–68, 84–85, 108–109, 224, 229, 232, 262

stock prices, 36, 55, 65, 66, 78, *80*, 92, 94, *94*, 95, *96*, 105, 110, 122, *126*, *148*, 150, *154*, 161, 164, *164–165*, *183*, *211*, 226, 229, 230, 232, 233, 238, 268, 306, 307, 309, 319, 320, 326, 327, 328, 330–331

stock-warrant financing, 320, 321

Stump, William, 63

subordinated debentures, 84, 91, 92–94, *108–109*, 233, 234, 238, 312, 313

subordinated debt, 84, 308, 311

success, preparing for, 289–290

Sun Microsystems, 317

Sylvan Learning Systems, Inc., 198

synthetic convertible, 103, 328

T

Taiwan, 278

tariffs, 46, 48, 61–62

TAT-G-1, 159

TAT-X link, 159

tax considerations for investors, 66

Taylor, Gerald H. (Jerry), xii, xix, 1, 2, 5, 23–25, 26, 33, 49, 56, 69, 70, 71, 74, 82, 97, 119–120, 127, 129, 141, 142, 150–152, 153, 154, 168, 192, 199, 205, *225*, 249, 251, 253–254, 260, 283, 284, 285, 286, 291–292, 295

TCI, 90

TDS, 127

teams, 206, 299

technology, xvi, xviii, 97–101, 168–170, 181, 258–259, 304, 306

Telechoice, 182

Telecom Italia, *201*

Telecom*USA, 160–161, 241

Tele-Communications, 183, 218

Telecommunications Act of 1984 (Great Britain), 204

Telecommunications Act of 1996, xii, 79, 184, 191–192, 193, 199, 219, 223, 245, 270, 313

telecommunications industry, xv–xix, 100, 206, 207, 215–223

telecommunications policy, 47, 191–193. *See also* regulatory issues

Teledesic, 219

Telefonica de España SA, *201*

Telefonica Panamerica MCI, *201*

Telefonos de Mexico, 158–159, 241

telegraph, 215

telemarketing, 149, 150, 152, 202, 286, 287, 288

telephone, early history of, 215–216, *221–222*

teleprinter service, 44

television networks, 71

telex service, 80–81, 83, 101, 139, 157, 158, 240

Telmex, xviii

Tesoro Petroleum, *320*

Texas East (regional carrier), *24*

Texas Instruments, 32

Texas Pacific (regional carrier), *24*

Thatcher, Margaret, 102–103

Thatcher, Simpson, 339

Thompson II, Josiah V., 27

Time, 117

Time Warner, 90, 218

Tisa, *201*

Todd, Mike, 8

Todd-AO, 8

toll-free 800 service, 136–137, 140, 171–172, 195, 202, 239, 243

"Tonight Show," 155

touch-tone telephone, 45, 114

Toyota, 11

Trans World Airlines, 58, 59
transistors, 216
transmission, 202–205
trends, 305–306
trial orders, 300
"True USA" campaign, 157
Turner Broadcasting, 90, 307

U

UBS Securities, Inc., 195
Uhl, Richard, 57
Ultrasonic Corporation of America. *See* Powertron
Unisource Consortium, *200*
United Airlines, 137
United Kingdom, xv, *200, 201,* 204, 263
U.S. Department of Commerce, 266
U.S. Immigration and Naturalization Service, 137
U.S. Interstate Commerce Commission, 266
U.S. Patent Office, 266
United States Postal Service, 147
U.S. Servicator, 11
U.S. Sprint, 62, 116, 117, 130, 137, 148, 150, 159, *161,* 172, 183, 188, *200, 201,* 217, 218, *241,* 275, 287
U.S. Telephone, *320*
U.S. West, 118, 218
University Communications council, 25
University of Mississippi, 212
USA Today, 168, 182, 281
UUNET, xiii, xv

V

Vail, Theodore, 29, 216
valuation ratio, xvii
VAULT architecture, 143
vendors, 33–35, 310, 312, 314–316, 338
venture capital, xviii, *317*

Vetere, Barry, 72
Viacom, 90, 307
video services, 178, 193, 197, 198, 263
voice mail, xix, 160, 195
Volvo, 72

W

Wall Street Journal, 147
warrants, 55–56, 67, 68, 103, 107, *108–109,* 110–111, 229, 232, 328–329. *See also* bond-warrant financing, stock-warrant financing
Washington Post, 52, 268
Washington Times, 131
WATS, 101
WATSbox, 44, 54
Western division, 119
Western Electric, 15, 38, 113, 186, 187, 216, 258
Western Tele-Communications, Inc. (WTCI), 63–64, 231
Western Union, 215–216, 217, *220,* 221, 222
Western Union International, 46, 81–82, 93, 127, 157, 234
Westinghouse, 46
Williams Telecommunications Company, 134
Wilson, Gov. Pete, xviii
Windows 95, 203
Wired, 180, 305
Wired Ventures Inc., 305
wireless communications, 51, 128, 191, 199, 217–218, 223
World Trade Organization (WTO), xvi, *200*
WorldCom, Inc., x, xii–xv, 161, *201,* 206, 210–212, 246
WorldPartners, *200*
Worthington, John R., 11, 12, *12,* 21, 43, 243, 250
Wright, V. Orville, 52–54, *53,* 80, 100, 101, 119, *123,* 129, 131, 132, 138, *229,* 250, 260, 285

XYZ

Xerox, 52, 53, 81, 82, 83, 229, 234
yield spreads, 87–88, 333–334
Young Presidents Organization, 196
zero-coupon notes, 307

List of Illustrations

Access charges as percent of revenue, 1984 vs. 1988, 136

Advertisement, 147

Advertisement for Execunet, 48, 49

Advertisement for networkMCI, 166, 178, 202, 244

Airsignal pager, 129

American Express card, 74

Angela Dunlap, 131

April 1981 financing decision, 92

AT&T advertising spending in 1980, 71

AT&T divestiture, 116

Baker Library, Harvard Business School, 6

Bert Roberts, 46, 226, 242

Bill Conway, 84

Bill McGowan, ix, 13, 15, 21, 27, 41, 161, 224, 227, 239, 243

Bill McGowan and Jack Goeken, 21

Bill McGowan and MCI Board member Alexander Buchan meet with COMSAT and Lockheed officials, 31

Bill McGowan, advertisement, fiber optic cable, 146

Bill McGowan, Nathan Kantor, and Edward Gallagher, 81

Billing, 203

Billion Dollar Day, 171

Burt Lancaster in MCI ad, 75

Capitol building, 43

Celebrating the FCC 1969 MICOM decision: Bill McGowan, Josiah V. Thompson II, Alexander Buchan, and Jack Goeken, 27

Cellular phone, 127, 245

Check for $988 million, proceeds of the 1983 bond-warrant financing, 106, 236

Children at computer, 77, 244

Communications network, 1974, 48

Comparison of MCI and Netscape initial public offerings, 36

Construction crew, meeting, microwave dish, 16

Convertible bonds, 96

Corporate Office: Bert Roberts, Bill McGowan, Orville Wright, and Richard Liebhaber, 123

Cover of the October 6, 1987 Financial World issue in which MCI was named one of the ten worst-managed companies in the nation, 133

Customer service, 203

Customer service representative, 67

Documents from the billion-dollar bond-warrant financing, 104

Early operators at work in the Bell system, 29

Employees, 140, 141, 209, 239, 244

Employees, fiber optic cable installation, 248

Employees, fiber optic cable, billing, 190

Entering the Canadian market, 1983, 102

Equal access voting, 113

Exterior of FCC building, 54

Faces of Milken Cube—perspectives key to a successful financing strategy, 303–306

Fiber optic cable and installation, 97, 98, 134, 202, 235, 239, 240, 242

Fiber optic cable, satellite dishes, 76

Financial offerings in first quarter of 1983, 320

Financing events, 1972–1983, 108–109

Fred M. Briggs, 194

Global network, 158, 240

High-yield market, 1980–1995, 87

Initial public offerings, 1972, 37

Institutional investors, 1995, 332

International telecommunications players, 200–201

Jack Goeken, 13, 18, 21, 27, 28, 51, 224

Jerry Taylor, 26, 225

John R. Worthington, 12

Judge Harold Greene, 79

Kenneth A. Cox, 42

L.A. Times headline: "$1.8 Billion AT&T Defeat," June 14, 1980, 78

Larry Harris, 32

Long distance prices, 1984 and 1996, 182

Long-distance market share, MCI and AT&T, 1975–1995, 186

Long-distance market size and MCI market share, 1984–1988, 138

Long-distance price comparison, 1984 and 1996, in residential and business markets, 182

Los Angeles Times article on Bill McGowan's death, 162

Map of MCI regional units, 1985, 117

Market capitalization, net income, revenues, and employees, 1974–1979, 65

Market share breakdown before merger with Telecom*USA, 161, 241

Market share, top three long-distance companies, 1985, 116

MCI ad, 147

MCI capital structure, 1975–1982, 86

MCI digital network, 1987, 135

MCI headquarters in Washington, D.C., 167

MCI logos over the years, 192–193

MCI system in 1978 after WTCI acquisition, 64

MCI under Bill McGowan, 1972–1991: employees, revenues, share price, and market share, 164–165

Michael Bader, 19

MCI: Failure Is Not an Option

Michael Milken, 90
Michael Rowny, 196
Michael Salsbury, 184
MICOM's sixteen regional carriers, 24
Microwave dishes, 30, 33
Microwave network construction, 14, 39
Mother's Day ad, 1980, 73
Net income, 1986, 132
New York Times article on AT&T divestiture, 114–115, 236
1-800-COLLECT, 175
Orville Wright, 53, 229
Pictures from the 1980s: Airsignal pager, satellite dish, AT&T out-of-court settlement, 238
Pictures from the dawn of competition: Tom Carter (developer of the Carterfone), the FCC building, cellular phone, chart of the AT&T divestiture, signing of the Telecommunications Act of 1996, first MCI logo, 223
Pictures from the dawn of telecommunications: Alexander Graham Bell's patent for telephone technology, Thomas Edison, launch of commercial telephone service, Guglielmo Marconi, 221
Pictures from the dawn of telecommunications: Samuel F. B. Morse, Western Union telegraph operators, early telegraph machine, first public telegram sent, Pony Express, 220
Pictures from the early days of MCI: Dallas One switcher, Orville Wright, 229
Pictures from the early days of MCI: Doug Maine, U.S. Court of Appeals, West Coast links, 231
Pictures from the early days of MCI: Execunet dialer, satellite, 228
Pictures from the early days of MCI: Jack Goeken, Bill McGowan, microwave dishes, 224
Pictures from the early days of MCI: Jerry Taylor, satellite in orbit, FCC meeting room, FCC logo, 225
Pictures from the early days of MCI: Telemarketer, network construction, Bill McGowan, 227
Pictures from the early days of MCI: Wayne English, advertisement for Execunet, 230
Pictures from the early development of telecommunications: J. P. Morgan, Congressional building, FCC logo, 222
Pictures from the early days of MCI: Chicago and St. Louis skylines, stock market, Bert Roberts, 226
President Bill Clinton and Al Gore at the signing of the Telecommunications Act of 1996, 191, 245
"Proof Positive" ad campaign, 172, 242
Railroad tracks, 3
Railroad tracks, Baker Library at Harvard Business School, microwave network construction, 2
Regional units, 237
Representation of Chicago-St. Louis link, 17
Revenue comparison to AT&T, 1975, 49
Revenue comparison to AT&T, 1979, 67
Revenue comparison to AT&T, 1980, 78
Revenue comparison to AT&T, 1982, 97
Revenue comparison to AT&T, 1984, 119
Revenue comparison to AT&T, 1990, 142
Revenue comparison to RBOCs, 1984, 118
Revenue comparisons to IBM (1985) and AT&T (1985), 125
Revenues, 1975–1996, 187
Revenues, net income, and costs of construction, 1968–1974, 45
Satellite dish, pager, employees, 112
Satellite dishes, 100, 122, 235, 245
Satellite in orbit, 99
Space shuttle launch, viii
Spoof of AT&T long-distance commercials, 70, 233, 272
Stock price and market share, 1990–1993, 154
Stock price reaction to antitrust decision, 1980, 80
Stock prices, 1973–1975, 55
Stock prices, 1975–1982, 94
Stock prices, 1982–1984, 110
Stock prices, 1985, 126
Stock prices, 1997, 211
Stock prices and net income 1975–1980, 59
Stock prices and revenue growth, 1989–1990, 148
Stock prices vs. S&P 500 average, 1975–1996, 183
Supreme Court, 60
Supreme Court building, 1980s TAC dialer, and customer service representative, 40
TAC dialer, 43
Tim Price, 143, 237
Tom Leming, 38
Video phone, 170
Vinton Cerf, 180
Wayne English, 58, 230
Xerox Company headquarters sign, 234

Index

Glossary

ADJUSTABLE RATE Interest rate or dividend that changes periodically. The adjustments are based on a standard market index, such as the rate on Treasury bonds or notes. A "floor" puts a limit on how low the interest can fall; a "ceiling" limits the rise. An adjustable-rate note pays a fixed interest rate for a set period, after which the issuer may change the rate. These notes typically range in maturity from five to ten years. The interest rate for adjustable-rate preferred (ARP) stock may be tied to the prime rate.

AMORTIZATION Reducing the cost value of a limited life or intangible asset through periodic charges to income. It can refer to depreciation, depletion, and the write-off of intangibles like goodwill, in each case gradually reducing the cost basis of assets through regular charges to income. In bond or preferred-stock investments, it's apportioning over the life of the investment any premium you may have paid over par value. In corporate financing, it's repaying a loan in installments that will cover the entire principal and interest by maturity. A company's amortization practices are detailed in its annual report in the notes to the financial statements.

ANALOG Representing data by physical variables, the way time is shown by the hands on a clock. A telephone turns voice vibrations into (analog) electrical vibrations that can be transmitted over phone lines. An analog signal is continuous, as opposed to a digital signal, which is broken up into numbers.

ANTITRUST Acting to prevent the emergence of monopolies or eliminate them if they do develop. The three key pieces of antitrust legislation are the Sherman Act of 1890, the Clayton Act of 1914, and the Federal Trade Commission Act of 1914. In addition, the Hart-Scott-Rodino Act of 1976 requires you to notify both the FTC and the Justice Department if you buy $15 million in a company's stock, or enough to own 15 percent of the company.

ARBITRAGE Buying a security, currency, or commodity on one market and simultaneously selling it (or an equivalent) on another market to take advantage of price differences. A simple arbitrage deal might involve buying a silver contract in Chicago for $4.70 an ounce and selling an equal contract in Hong Kong for $4.72. A more complicated arbitrage strategy might involve several variables such as currency rates, interest rates, and derivatives like options. An arbitrageur, for example, may buy a stock with French francs in Paris while selling an option on it in U.S. dollars in New York. Because markets tend toward equal prices, arbitrageurs must act quickly to take advantage of temporary differentials. Also, because the price differences are small, arbitrageurs need to trade in big blocks to reap sizable profits.

ASYNCHRONOUS TRANSFER MODE (ATM) A way of transmitting signals that was developed to provide flexible and efficient service for compressed video and other applications. It involves sending variable bits of data.

BABY BELLS Seven regional Bell operating companies spun off by AT&T to comply with the 1982 AT&T antitrust settlement. See also "REGIONAL BELL OPERATING COMPANIES."

BACKBONE NETWORK A transmission facility designed to connect lower-speed distribution networks, channels, or clusters of communications devices or terminals.

BANDWIDTH Measure of the capacity of a channel to carry communications. Whereas voice transmission requires little bandwidth, video transmission requires a great deal.

BELL OPERATING COMPANY (BOC) One of 24 local telephone companies that were part of the Bell System before divestiture. Seven regional Bell holding companies now own and manage all but two of the BOCs.

BETA Measure of a security's or portfolio's volatility (riskiness) relative to all other securities in the market (e.g., S&P 500, Wilshire 5000, etc.). Beta is calculated by comparing a security's historical price movements to the movements of the overall market (i.e., standard deviation).

Stock prices change in response to changes in underlying factors such as interest rates, inflation, and investor confidence. When a single stock has a standard deviation greater than the overall market, the stock has a beta greater than 1.0. When a single stock has a standard deviation lower than the overall market, the stock has a beta less than one. When a stock has a standard deviation equal to the overall market, the stock has a beta equal to 1.0.

If the market rises 10%, a stock with a beta of 2.0 is expected to rise 20%. When the market falls 10%, a stock with a beta of .5 is expected to fall 5%. If the market rises 10%, a stock with a beta of 1.0 is expected to rise 10%.

Beta measures systematic, or market, risk. Systematic risk is the risk all stocks face from changes in the underlying market factors listed above. Because all stock face systematic risk, it cannot be eliminated through diversification.

BIT A binary digit—the most fundamental form of digital signal. It's a condition that represents one of two digits, either 1 or 0. Physically, it can be as simple as a magnetic spot on a disk or a pulse of electricity moving through a circuit.

BIT RATE The number of bits per second that a channel can transmit.

BOND Long-term (>10 years) debt security issued by a corporation or government entity, usually in multiples of $1,000. A bond obligates the issuer to pay, usually twice yearly, a fixed amount of interest at specific intervals, and repay to bondholders the principal of the loan at maturity. Since bondholders are creditors, not shareholders, they have no ownership privileges, such as voting. However, when a firm is in distress bondholders have a senior claim to a firm's assets over stockholders.

BOND RATING Evaluating bonds based on their risk of default. Also called quality ratings. Several organizations publish bond ratings, most prominently Moody's Investors Service and Standard & Poor's Corporation. Bonds considered high grade, or investment grade, have ratings of Aaa through Baa3 in the Moody's system; in the S&P rankings, they have AAA through BBB-. Bonds with ratings below those are considered speculative, or high-yield bonds.

BOND/WARRANT UNIT A security in which a bond and a set number of equity warrants (options to buy stock at a fixed price) are sold as a unit, sometimes called "synthetic" converts. In accounting for these units, the issuer records the bond as debt and the warrants as equity. In contrast, traditional convertible bonds are recorded entirely as debt until exchanged. If desired, the bond portion is "usable as currency"—generally at par—to exercise the warrants. Unlike conventional convertible securities, the debt (bond) and equity (warrants) can be traded separately.

BREAKEVEN TIME The period needed for a company to recoup its initial investment or for sales revenues to cover production costs. At the breakeven point, there's neither a profit nor a loss. In reference to convertible bonds, it's the period necessary for cumulative interest on the bond to equal the conversion premium.

BUS In digital systems, a bus, or highway, is one or more conductors that connect a related group of devices. A bus may connect a computer processor and memory to units that control communications, a disk, a printer, and so on. A bus always connects several devices in a grouping.

BUSTED CONVERT A convertible that trades like a straight bond because the market price of the common stock has fallen so low that the conversion feature is valueless. Assuming the issuer can continue to pay interest, a busted convert can make an attractive investment. The stock price may rise to a point where the conversion feature is valuable.

CALL FEATURE Provision of a bond agreement that allows the issuer to repurchase the bond before it matures. The call price of a bond—the price the issuer pays to redeem it—is usually higher than the bond's face value (typically $1,000). The difference between the call price and the bond's face value is the call premium. As bonds get closer to maturity, the call premium declines. Companies or other bond issuers may exercise their call features when interest rates have fallen. They repurchase the high-interest bonds and issue new ones at lower interest rates. In the mid-1990s, the number of bonds issued with call features declined sharply as institutional investors like mutual funds demanded higher call premiums.

CAPITAL STRUCTURE A corporation's common stock, preferred stock, other securities, and debt. Also called capitalization. Usually excludes current liabilities like accounts payable and short-term debt.

COAXIAL CABLE An insulated conductor surrounded by a second, cylindrical conductor and an insulating sheath. The outer conductor often consists of copper braid or tubing.

COMMON STOCK Common stock is a share in a company's ownership. Holders of the stock supply equity capital, often have voting rights, and elect the board of directors. This allows majority shareholders to indirectly control the company. Shareholders sometimes benefit from a company's success through income (the receipt of dividends) and by capital appreciation (increase in stock price). Common shareholders have only a residual claim to the assets of the company—behind creditors, debtholders, the IRS, and preferred shareholders—should it declare bankruptcy. Therefore common stockholders are the last in line to collect funds in case of bankruptcy.

COMPETITIVE ACCESS PROVIDER (CAP) One of the companies that, since the AT&T divestiture, can compete for long-distance business on an equal footing.

CONVERGENCE The merging of the computer, telephone, cable television, and entertainment industries. It's a result of the overlap of technologies and markets. Convergence has been accelerated recently by a steady stream of mergers and acquisitions across industry lines and by a loosening of regulatory restrictions.

CONVERSION PREMIUM Amount by which the market price of a convertible security exceeds its conversion value. When the conversion premium is high the convertible trades like a bond. When there is no conversion premium (the price of the convertible has reached conversion parity) the convertible will rise along with the underlying common stock.

CONVERTIBLE BOND A convertible bond is typically an unsecured debt obligation of a company. It is exchangeable into a predetermined amount of common stock at a predetermined price in the future. In bankruptcy, these bonds would rank below subordinated senior debt and above preferred and common stock. In addition to scheduled interest payments, convertibles offer the investor equity participation in the event the price of the stock rises above the conversion price. These bonds are booked as debt, and may carry lower coupon payments than straight debt of similar seniority and credit quality.

CONVERTIBLE PREFERRED STOCK Stock that allows the holder to convert shares into another security, usually common stock. The preferred shareholder receives dividend income, while retaining the opportunity to gain from stock appreciation. For issuing corporations, dividend payments, unlike interest payments, are not tax-deductible. But corporations receiving dividends can exclude 70 percent of the preferred dividends from their taxable income.

COUPON A bond's stated interest payment. It's the amount paid on the face value, or par value, of the bond. If you buy the bond for less than par value, your actual yield will be higher than the coupon rate. If you pay more than face value, the yield is lower than the coupon rate. Most bonds are registered in holders' names, and interest payments are sent to the registered holder, but the term coupon rate is still widely used. Interest is typically paid twice a year. Example: A bond with a $1,000 face value has a 9 percent coupon. It pays the bondholder $90 a year in two $45 installments.

COVENANT A promise to perform or refrain from performing certain acts. Debt covenants are found in the agreement, known as the indenture, filed with the Securities and Exchange Commission at the time of a debt offering. Typical covenants include restrictions on borrowing over a certain amount or from letting liquidity or operating ratios cross certain thresholds.

DEBENTURE Unsecured bond. With no collateral pledged specifically to the bondholder, debentures stand in line behind secured debt. Due to increased risk debentures pay more interest than secured bonds, and they are by far the most common form of corporate bond, usually issued by large companies that have strong credit records. Debentures often have conversion features or warrants that allow holders to exchange the bonds or warrants for common stock on a given date. The bondholder may also have the option of converting the debenture to common stock if the company defaults on interest payments or bond redemption. An indenture, a document filed with the Securities and Exchange Commission, details all terms of the bond.

DERIVATIVE SECURITIES Financial instruments with value based on (derived from) one or more other securities or indexes. Examples range from stock options trading on major exchanges to complex private deals involving different currencies and interest rates.

Many businesses use derivatives to reduce risk. Hedging insulates a company from changes in interest or currency-exchange rates. As derivatives have become more popular, however, companies have also begun trading them in more speculative ways to profit from swings in interest rates or the prices of commodities or currencies.

Derivatives are traded two ways. One is on exchanges, where hedgers and speculators buy and sell standardized contracts through licensed brokers, backed by a centralized clearinghouse that requires minimum deposits, or margins, to cover the value of the contract. The second way derivatives are traded is in over-the-counter (OTC) transactions privately negotiated between two or more parties, often using a major bank as mediator. OTC deals grew about eightfold from the mid-1980s to the mid-1990s.

In the 1980s, the use of derivatives mushroomed along with the growth of both international trade and the near-instantaneous movement of capital through electronic transactions. In the mid-1990s, the value of the securities underlying derivatives was estimated at greater than the value of all the world's stocks and bonds (roughly $32 trillion).

DIRECT BROADCAST SATELLITE (DBS) Broadcasting television signals directly from a satellite to 18-inch-diameter roof-mounted dishes on homes. DBS is unlike original satellite broadcasting in that the dishes are far smaller and the signal is digital, not analog, allowing broadcast of hundreds of channels. Digital transmission provides much better sound quality than does analog.

DOW JONES INDUSTRIAL AVERAGE (DJIA) Oldest and most widely followed U.S. stock market index. The DJIA is a price-weighted average of 30 widely traded stocks on the New York Stock Exchange (NYSE). The companies change from time to time, but they usually represent between 15 percent and 20 percent of the market value of all actively traded stocks on the NYSE. The DJIA is adjusted for the substitutions, mergers, stock dividends, and splits that have occurred since it was first published in 1896. Because of 100 years of adjustments, it's value is greater in measuring price movements than absolute price levels.

DUOPOLY Exclusive control by two companies of a market for particular products or services. A duopoly may arise from legal protection, proprietary technology, command of supply, control of distribution, and so on. See also "MONOPOLY."

EQUAL ACCESS Provision in the 1982 AT&T antitrust settlement that gave all long-distance carriers equal access to local networks. Thus, all long-distance customers had access to the carrier of their choice through their local telephone company. Instead of dialing 22 numbers to reach someone in a distant city, callers could just dial "1," the area code, and the 7-digit phone number.

EQUITY KICKER Attaching an offer of an ownership position to a loan or debt instrument. You might offer lenders an opportunity for an equity stake in your business to reduce the interest rate or to improve other terms of your loan. Equity kickers attached to bonds include warrants, rights, and options. The convertibility provisions of convertible securities are also equity kickers. In other loan agreements, a borrower may offer a lender a small ownership position in the project or acquisition that's being financed. In that case, when the property is sold, the lender is likely to get additional income.

EUROBOND CONVERTIBLE SECURITY Convertible corporate bond denominated in one country's currency and issued in another. The term would apply, for example, to a U.S. company selling dollar-denominated convertible bonds in Europe or even in Japan. The corporation's home may be in yet another country, so a Brazilian company might issue dollar-denominated bonds in London. Eurobond convertibles usually can be exchanged for the issuer's stock and include a put, which gives the holder the right to demand repayment by a specific deadline.

EXCHANGE CARRIER Any company, regardless of whether it was once one of the Bell carriers, that provides telecommunications service within a "LOCAL ACCESS AND TRANSPORT AREA (LATA)."

EXECUNET Intercity telephone service launched by MCI Communications in the early 1970s. It also refers to a decision by the Federal Communications Commission, eventually reversed by the courts, to shut down the Execunet service.

FEDERAL COMMUNICATIONS COMMISSION (FCC) Federal agency created in 1934 to oversee the nation's interstate and international communications. These now include telephone, telegraph, radio and television broadcasting, cable television, and satellites.

FIBER-OPTIC CABLE Precisely aligned bundle of optical fibers that transmits data from one end to the other. Preferred over copper wiring for telecommunications because of its much higher capacity, immunity to electrical interference, and reduced weakening of the signal.

FISCAL YEAR (FY) The 12-month accounting period a company uses. For some two-thirds of U.S. publicly traded companies, this is the same as the calendar year. But it makes sense for some companies to use a different fiscal year. Department stores are the most common example. Many of them use a fiscal year that runs through January, when the holiday season is behind them. A number of other industries also use fiscal years that fit their business. Example: Education Alternatives, a company that manages schools and consults on education, uses a fiscal year that ends June 30, along with the school year.

FORCING CONVERSION When a convertible security is called in by its issuer. Despite the name, holders of the security aren't forced to convert it to the underlying common stock. They can accept the issuer's call price or sell the security on the open market during the call's notice period. If the common stock price is higher than the conversion price, most people will take the conversion option. The optimal time for a company to call its convertibles is when the conversion price equals the underlying stock's market price. But that rule is largely theoretical. Most forced conversions come when the stock price goes way above the conversion price.

FREQUENCY The number of cycles per second in alternating current. One cycle per second is a hertz (Hz). A kilohertz (kHz) is a thousand cycles per second.

GRAHAM-WILLIS ACT Act of U.S. Congress in 1921 that deemed AT&T a natural monopoly. A natural monopoly is a business that requires such economies of scale to operate profitably that only one company is economically viable to serve the industry. AT&T's monopoly status, however, has fallen to legal assaults, many of them mounted by MCI. See also "MCI DECISION."

GREEN SHOE Clause in an underwriting that allows additional shares to be sold during a public offering after the initial allotment runs out. Usually the issuer agrees to sell a specific number of added shares if the underwriter believes the issue will be oversubscribed and the issue price is lower than the price at which the stock could trade. The term comes from the Green Shoe Company, which first offered such an arrangement to its underwriter.

HIGH DEFINITION TELEVISION (HDTV) Television broadcasting standard that provides much higher picture, color, and sound quality than conventional television. The HDTV standard includes resolution of 1,150 scanlines, digital sound, and wide-screen format.

INDEPENDENT TELEPHONE COMPANY (ITC) A local exchange carrier that is not one of the 22 Bell operating companies (BOCs). ITCs generally do not come under the terms of the Modified Final Judgment that divested the BOCs from AT&T, but the larger ones are bound by separate consent decrees.

INFORMATION SUPERHIGHWAY Futuristic concept of a telecommunications network that links millions of computers and related devices around the world. As conceived, the superhighway would enable widespread access to unlimited information and on-line transaction of business in every sphere of life, from commerce and education to politics and entertainment. Futurists often point to the Internet as a hint of the superhighway's potential.

INITIAL PUBLIC OFFERING (IPO) First offering to the public of a company's common stock. After the IPO, the publicly held shares may be traded on a stock exchange or the over-the-counter (OTC) market. A company may go public when it needs more capital to support or expand operations. Or the company's private owners may want to sell part of their stake. The existing private investors may see their holdings increase sharply in value if the market sees the company's potential for growth. Companies selling shares to the public must register them and meet disclosure requirements of the Securities and Exchange Commission (SEC).

INTEGRATED SERVICES DIGITAL NETWORK (ISDN) High-speed telephone link that allows the transmission of voice and data over the same wire. ISDN transmits voice and data with a digitized signal four times faster than today's highest modem speeds. ISDN may also transmit video but requires a fiber-optic infrastructure for high resolution. Despite ISDN's promise, and the costly upgrading already performed by telephone companies in half of U.S. cities to accommodate the technology, experts believe that fiber-optic technology will displace ISDN as the technology for carrying voice, data, fax, cable TV, and video.

INTERNATIONAL RECORD CARRIERS (IRCs) An elite international club whose membership included ITT, Western Union International, FTC Communications, TRT Communications, and Cable & Western, Ltd. These IRCs were the only companies authorized to send and receive telexes between the United States and foreign countries. Before it could enter the international market, MCI had to acquire an IRC.

INTRA-LATA (INTRA-LOCAL ACCESS AND TRANSPORT AREA) Refers to telecommunications services, revenues, and functions confined to a single LATA. Equivalent to the span of operations of a local carrier.

LIGHTWAVE SYSTEM An optical system that uses optical fibers to transmit light pulses.

LOCAL ACCESS AND TRANSPORT AREA (LATA) Service area for one of the 22 Bell operating companies. These LATAs, defined in the Modified Final Judgment that broke up AT&T, were intended to distinguish between local and long-distance calling markets.

LOCAL AREA NETWORK (LAN) System of linked computers and peripherals like printers in an office or building or in a small geographic area. A LAN connects computers via a cable that allows users to share information and software. A wide area network is larger and covers more geography, using dedicated telephone lines or radio waves. The most common way for personal computers to work together is in a client/server network where one computer (the client) gets many of its applications and files from another (the server).

LOCAL EXCHANGE CARRIERS A company that provides telecommunications service within a local access and transport area (LATA).

LOCAL LOOP A transmission path between a customer's premises and a central office.

LONDON INTERBANK OFFERED RATE (LIBOR) Interest rate international banks usually charge each other for large Eurodollar loans. Like the prime rate for dollar loans in the United States, LIBOR is also a base rate for Eurodollar loans to the international banks' customers. A developing third world country, for example, may have to pay two points over LIBOR when it takes out a loan in Eurodollars.

LYON See "LIQUID YIELD OPTION NOTES."

M1/M2/M3 Aggregate measures of money supply. M1 is the yardstick U.S. economists usually use to measure the money supply. It includes all the currency in circulation (outside of bank vaults) plus traveler's checks, checking accounts, and other accounts that can be transferred to checking accounts instantly and entirely. So a savings account linked to a checking account is part of M1, but a money market account is not because its check-writing privileges are limited.

A somewhat broader version is M2. It boosts the total by including money market accounts, other savings accounts, and time deposits (up to $100,000) that aren't included in M1. These money market and savings accounts are called near moneys.

M3 includes M2 plus most additional forms of savings. One advantage of looking at M2 and M3 is that they tend to be more stable than M1. When banks are offering high interest rates on checking accounts, people shift funds from their money market and savings accounts to checking. That has no effect on M2 or M3, but M1 will show a sharp rise.

MARKET ADJUSTED DEBT (MAD) RATIO Debt, adjusted to its current market value, divided by the sum of adjusted debt and equity adjusted to market value. Companies with lower ratios are perceived to be of better quality in the marketplace.

MCI DECISION Decision by the Federal Communications Commission on August 12, 1969, to approve the construction applications of MCI. The decision, passed by FCC commissioners by a 4-3 vote, was the beginning of the end of AT&T's almost 70-year phone monopoly. The decision earned Jack Goeken the label "Giant Killer."

MICOM Company organized in 1968 to commercially exploit pioneering work in the field of common-carrier microwave communication systems. It later became MCI and, under the leadership of William (Bill) McGowan, pursued a vision of creating a nationwide network of regional microwave companies.

MICROWAVE Transmission in the electromagnetic spectrum above 1 Ghz. Microwave transmission frequencies are used in satellite communications and earth-based line-of-site systems.

MICROWAVE COMMUNICATIONS SYSTEM System of communications stations for transmitting short electromagnetic waves. In telephone communications, a microwave relay station is necessary every 26 miles.

MOBILE SATELLITE STATION (MSS) Movable device for receiving satellite signals. These include devices used in the marine, rail, and trucking industries for navigation and positioning.

MODEM (MODULATOR/DEMODULATOR) Device that allows computers and terminals to communicate over phone lines. Modems transform the digital pulses generated by a computer or terminal into analog signals for transmission and then convert the analog signals back to digital pulses understood by the receiving computer or terminal.

MODIFIED FINAL JUDGMENT Ruling issued by U.S. District Judge Harold Greene in 1982 that ordered the 1984 breakup of AT&T. Judge Greene's ruling concluded a Justice Department antitrust suit against AT&T by modifying a 1956 consent-decree judgment. The 1982 judgment ordered AT&T to spin off the seven regional Bell operating companies.

MONOPOLY Exclusive control of a market for a particular product or service. The control can stem from factors like legal protection, barriers to entry, economies of scale, control over required inputs, proprietary technology, control over distribution, etc. Because of the lack of competition, monopolized markets often lead to high prices and slow responses to customer needs. The Sherman Act of 1890 prohibits conspiracies and combinations in trade that lead to monopolies. The Clayton Act and the Federal Trade Commission Act, both passed in 1914, prohibit many monopolistic practices.

MULTIPLEXING Allows for more efficient use of transmission capacity by combining a number of signals in a single circuit. The two basic methods used in telecommunications systems involve either dividing frequencies or slicing time to send signals over a single line.

MUNICIPAL DEBENTURE Bond issued by a state or city government, or a government agency. Sometimes called a muni. These bonds raise funds for specific projects or for general needs. Interest paid on a muni is usually exempt from federal income tax and from taxes in the state that issues it.

NARROWBAND Telecommunications lines or circuits that carry data at speeds of 2,400 bits per second or less. It's also sometimes used to designate channels able to carry data at speeds of 200 bits per second or less. In cellular radio, it refers to splitting FM channels to gain more channels and greater capacity.

NASDAQ Computerized trading system serving as the primary over-the-counter (OTC) securities market in the United States. The acronym stands for the National Association of Securities Dealers Automated Quotations System.

The OTC market was for start-up companies and others too small to meet the listing requirements of physical markets like the New York Stock Exchange (NYSE) and American Stock Exchange (Amex). But lately a number of companies have stayed on NASDAQ long after they qualified for exchange listing. These include such giants as Apple Computer and Microsoft Corporation.

In the volume of shares traded, NASDAQ is hot on the heels of the U.S. market leader, the NYSE.

NASDAQ collects bid-and-ask prices from brokers around the country and disseminates them electronically. The brokers then trade with each other over the phone or electronically. As in physical markets, some brokers serve as market makers, buying and selling shares in specific companies to maintain orderly trading in those issues. NASDAQ also provides an alternative market for stocks listed on the major exchanges. Many daily newspapers publish NASDAQ stock quotes as well as those from physical markets.

NATURAL MONOPOLY Monopoly in an industry where economies of scale are so significant that only one very large firm can operate viably. The term often applies to utilities.

NETWORK HARM Damage to a telephone network. The term refers to past claims made by AT&T that non-AT&T devices connected to the network would cause damage.

NON-FACILITIES-BASED COMPETITION Competition for long-distance phone business by companies that buy or lease, rather than own, network capacity. Non-facilities-based competitors, which resell other firms' excess fiber-optic capacity, often offer specialized long-distance services to small businesses.

OPTICAL FIBER Guide for electromagnetic waves in the visible light and infrared spectrum. Cables can contain up to 136 fibers. Each one is composed of a central core of glass or plastic in which light waves travel surrounded by a light-absorbing jacket that prevents interference to or from the other fibers.

PERSONAL COMMUNICATION SERVICE (PCS) Wireless phone service that tracks customers' locations and routes incoming telephone calls to them. Cellular telephone carriers have offered a version of PCS for years. The U.S. government has auctioned an airwave spectrum for vastly expanded and less expensive PCSs, likely to rapidly accelerate growth of the wireless communications market.

PERSONAL COMMUNICATIONS Broad range of equipment and systems that connect human operators with each other and with telecommunications and information services. Cellular phones, personal communicators, radio pagers, and remote data entry terminals are all personal communication devices. The emerging class of personal communications services (such as Motorola's Envoy) will integrate phone, fax, messaging, and data storage in a tiny transportable package.

PLAIN OLD TELEPHONE SERVICE (POTS) Basic telephone service, without added features such as third-party calling, call waiting, and call forwarding.

PREFERRED STOCK Security that, like common stock, shows ownership in a corporation. Dividends on preferred stock are typically paid in fixed amounts; they won't grow over time. Both common and preferred shares are forms of capital stock. Holders of these securities get paid after bondholders in the event of liquidation. Preferred stockholders, though, have priority over those who own common stock in dividend payments and in any liquidation of the company. As a rule, preferred stockholders don't have voting rights. Since preferred stock pays dividends in fixed amounts, it seems like a bond paying interest in fixed amounts. But dividends, unlike interest, aren't tax deductible to the corporation.

PRIME RATE Interest rate leading banks charge their best, most-creditworthy commercial customers. It's a benchmark for other loans. With a prime rate of 8 percent, for example, a company might negotiate a loan at prime plus two, or 10 percent. An especially well-financed company might even get a loan at prime minus a half, or 7.5 percent.

PRIVATE LINE SERVICE Any of a variety of specialized telephone services. For example, the "Series 11,000," which AT&T proposed in the 1960s, was a private line service that would have provided a continuous wideband spectrum for both data and voice.

PRIVATE NETWORK A network made up of circuits and switching equipment for the exclusive use of a single organization.

PROTOCOLS Procedures for starting, maintaining, and ending data connections. Protocols define the precise methods of data exchange, including means of system control, formats, and patterns of bits, data rates, and timing of signals.

REGIONAL BELL OPERATING COMPANIES (RBOCs) The seven regional phone companies divested by AT&T in compliance with a 1982 court order—NYNEX, Bell Atlantic, BellSouth, Ameritech, U S West, Southwestern Bell, and Pacific Telesis.

RULE 415 (SECURITIES AND EXCHANGE COMMISSION) SEC provision allowing a company to file a registration statement (known as a shelf registration) as much as two years before actually offering securities for public sale. The registration must be updated quarterly, but it can be used quickly when the company decides that market conditions are right for making the offering. See also "SHELF REGISTRATION."

SATELLITE COMMUNICATIONS Use of satellites to receive signals from earth, amplify them, and send them back to earth stations, downlinks, or personal satellite dishes. Satellites use high-frequency waves, or microwaves, to transmit their signals.

SECURITIES AND EXCHANGE COMMISSION (SEC) Federal agency overseeing all aspects of the public trading of securities. The Securities Exchange Act of 1934 created the SEC to regulate securities markets, including the flow of information to investors and potential investors. The SEC defers to the Financial Accounting Standards Board on the rules accountants must follow when preparing financial statements. But the SEC has a lot to say about exactly what information must be disclosed when and how, and the contents of its required documents, like annual reports, 10-Ks, and prospectuses. Even the listing of officers and directors usually found on one of the last pages of a company's annual report is mandated by the SEC.

SELL-SIDE ANALYSTS Refers to securities analysts in brokerage houses and investment banks that are in the business of selling securities to institutions and individuals. A sell-side analyst recommends securities for customers to buy. See also "BUY-SIDE."

SERVER In a network, a computer that holds applications and files for use by the other computers in the network, which are called clients. The servers may make databases, word processing, printing, video, or other services available to the client computers. A gateway is a server that gives client terminals access to other networks or information systems, such as the Internet.

SIGNAL In telecommunications, a pulse or frequency that can be varied to convey information.

SIGNALING In telecommunications, sending a telephone signal between the telephone customer and the telephone company. Signals include ringing and dialing.

SIGNALING EFFECT Theory that the price of a company's common stock tends to fall when the company announces that it will issue new shares and rise when it announces a new debt issue. The presumption is that management—which has better information than investors do—will sell its equity only when it's pessimistic about the future. If management is optimistic about the company's future—according to the theory—it will raise money by borrowing rather than give up any of the ownership in the company.

SINKING FUND A separate fund established by the borrower to either retire a portion of a bond issue or accumulate cash that is used to retire the bonds at maturity. Sinking fund provisions, detailed in the bond indenture, enhance the safety of the investment, since cash will be accumulated on a set schedule. This added degree of safety allows the issuer to offer bonds at a lower interest rate. Some preferred stock is also issued with sinking fund provisions.

SONET Form of fiber-optic telecommunications transmission that allows the transportation of digital signals with different capabilities. Initiated by the regional Bell operating companies as an alternative to asynchronous transmission, it makes possible the networking of transmission products from multiple vendors.

Glossary

SPECIALIZED COMMON CARRIER DECISION Decision by the Federal Communications Commission in 1971 that helped open the telecommunications market to competition. MCI Communications followed up the decision by launching its first private intercity microwave system for commercial telephone users.

STRIPPED YIELD Return on the debt portion of a bond/warrant unit after subtracting the value of the issued warrant.

STRIPS U.S. Treasury or municipal securities with separate interest and principal components. The holder can choose to separate the bond's interest and principal portions.

SUBORDINATED DEBENTURE Bond of the most junior form, unsecured by any collateral, and payable only after debtholders at all other levels have been paid. The more senior debt includes loans from banks or insurance companies and any bonds or notes not clearly designated as junior or subordinated.

SUPERVISORY SIGNALS In telecommunications, the signals used to indicate to a piece of equipment, an operator, or a user that a particular state in a call has been reached. These signals may simplify the required responses.

SWITCH A device that makes, breaks, or changes the connections in an electrical circuit. In the telephone business, a switch is a central switching office, which is found in every city or town.

SWITCHING In telecommunications, connecting a caller to a call recipient.

SYNTHETIC CONVERTIBLES Bond/warrant units where the bond can be used in whole or part as cash for the exercise of the warrants. Bonds that can be cashed this way are called usable bonds. They usually can be cashed in at full face value whenever the holder exercises the warrants. These bond/warrant units, taken together, resemble convertible bonds, with the notable exception that the warrants usually will expire before the bonds' maturity.

T1 CARRIER A multiplexed digital transmission facility capable of supporting 24 voice channels.

TARIFF (FCC) List of charges filed by a telephone company with the Federal Communications Commission (FCC) or a state public utility commission (PUC). The tariff includes prices for all services and equipment offered. The FCC and PUC accept the tariff until challenged in court. If a court strikes down the tariff, the phone company must refund customer money.

TELECOMMUNICATIONS A process enabling one or more users distant from each other to share information in a usable form instantaneously. The word is derived from tele, Greek for "far off," and communicate, Latin for "to share."

TELECOMMUNICATIONS NETWORK Series of interconnected facilities designed to carry signals from a variety of telecommunications services.

TELEPHONY Originally, the transmission of sound over long distances. The word has come to mean translating voice, data, video, and images into signals and transmitting them over long distances. With inventions such as the facsimile machine and data modem, telephony now includes the integration of the telephone and computer.

TELOCATOR DATA PROTOCOL (TDP) Way for pagers and personal computers to communicate with each other. It was developed in 1993 by Telocator, a trade group now called the Personal Communications Industry Association. The protocol allows alphanumeric data to be sent from computers to pagers at a low cost. So messages and such services as news and weather can be read directly from pagers.

TIME DIVISION MULTIPLEXING (TDM) Transmission strategy that uses time, rather than frequency, in order to accommodate multiple signals.

TRANSCEIVER Device that can both transmit and receive. A local area network (LAN) transceiver consists of a transmitter, receiver, power converter, collision detector, and jabber detector (a timing circuit that protects the LAN from a continuously transmitting terminal).

TRANSMISSION FACILITIES Communications paths that carry user and network control information between nodes in a network. Transmission facilities consist of a medium, electronic amplifiers, switches, and multiplexing equipment.

TRANSMISSION MEDIUM Any substance that can be used for propagation of signals, in the form of electromagnetic or acoustic waves, from one point to another.

TRANSPORT SERVICES Network transmission, switching, and related services that support information transfer capabilities between facilities.

TREASURY STOCK Issued stock that has been reacquired by the corporation. The company takes it out of circulation pending later retirement or resale. Because it's no longer outstanding, treasury stock carries no vote, bears no dividends, and is excluded in figuring per-share financial ratios. It may be used to cover options, warrants, and convertible securities.

TRUNK Communication path connecting two switching systems. A trunk is used to establish end-to-end connections between customers.

TWISTED PAIR Two insulated copper wires, twisted together. The twists are varied in length to reduce the potential for interference. Twenty-five pairs are bound together in a common cable sheath.

UNDERWRITING Agreement by investment bankers to buy for resale an entirely new issue of securities at a fixed price from the issuing company. The underwriters assume the risk of reselling the securities to the public and profit from the spread between the price they pay the company and the price they get in the market. Most such deals are managed by a lead underwriter, which assembles a syndicate of investment bankers to spread the risk.

VALUE-ADDED NETWORK SERVICES (VANS) Specialized services, offered over regular and special carrier networks, such as on-line databases, electronic yellow pages, videoconferencing, voice mail, video games, home shopping, and home banking.

VELOCITY The rate at which a nation's money supply turns over. Velocity is equal to GDP divided by the money supply.

VENDOR FINANCING Vendor financing is credit extended to a company by its suppliers, usually less than 10 years in term. Normally, it is senior secured debt, backed by equipment. In the event of a bankruptcy, because of its seniority in a company's capital structure, this type of security is paid off before subordinated debt, preferred stock, or common stock. This type of financing may be particularly effective when a company has numerous suppliers, which can be forced to compete not only on the products they provide but also on the credit terms they extend.

VERTICAL INTEGRATION The extent to which a company controls the successive stages in the supply of a product, from productivity raw materials to retail sale. Backward, or upstream, integration occurs when a company takes control of the stages toward the supply of raw materials. Forward, or downstream, integration occurs when a company takes control of the stages toward distribution to the consumer. Decisions over the degree and direction of vertical integration involve both financial and strategic considerations. If the expansion doesn't fit the company's core competencies, it might very well be a failure.

VIDEOCONFERENCING The real-time, two-way transmission of voice and video between two or more locations. Voice and video analog signals are digitized before transmission and then converted back to analog form at the receiving end. A variety of video compression techniques is employed.

VIDEO ON DEMAND Instant access by home viewer at any time to any movie or television show.

WARRANT Security allowing its holder to buy a number of common shares at a set price in the company issuing the warrant. For the investor, a warrant works like a call option, but any shares bought will come from the company rather than another investor. Warrants also usually have longer terms than options and occasionally have a perpetual life. Companies often issue warrants as sweeteners in the sale of a new issue of bonds or preferred stock. Warrants entitle holders to neither dividends nor voting rights.

WARRANT PREMIUM Price of a warrant minus its value when exchanged for the underlying common stock. This is what the investor is paying for the warrant's potential increase in value. In general, premiums tend to be higher on longer-life warrants. As warrants approach expiration, premiums shrink. Premiums also are affected by dividends on the common—the larger the dividend, the smaller the premium. That's because the dividends make it attractive to exercise the warrants. Expectations for the stock price also play a part in determining warrant premiums (as do supply and demand, plus whether the warrant is listed or not).

YIELD SPREAD Difference in yields between bonds of different quality. Also, it's the difference in yields between bonds and other securities. A large yield spread between stocks and bonds is called the yield gap.

YIELD TO MATURITY (YTM) Annualized rate of return on a bond given the market price, interest, redemption price, and time until maturity. YTM takes into account timing of cash flows and, if the bond was bought at a discount or premium, capital gains or losses.

SPURGE INK! OVERVIEW

Introducing
Spurge Ink!

Spurge Ink! is a business communications firm whose primary objectives are to empower consumers to take control of their own destiny—by helping them improve their careers, build their businesses, and invest their money. The company specializes in producing user-friendly information and delivers it through a wide variety of media including books, newsletters, seminars, videos, audio tapes, radio and television programming, as well as online. The company has won awards for its products, including the prestigious "50 Best Books of 1996" from the American Institute of Graphic Arts.

Publishing
- Books/Newsletters/Online
- Newspaper Columns/Magazine Articles

Media
- Radio/Television Programs
- Video/Audio/Interactive CD-ROMs

Training
- Investment/Career/Motivational
- Seminars/Presentations
- Educational/Business Products

Consulting
- Financial/Marketing/Management

Design
- Logo/Brand Image Development
- Informational/Marketing Materials

Making Business Accessible

Money Clips
365 Tips That Will Pay One Day at a Time

Lorraine Spurge

ISBN: 1-888232-44-7
256 pages 5 1/2 x 8 1/2
Business/Finance/Investment
$13.95 softcover (In Canada: $19.95)
Rights: World

From the popular radio and television mini-feature, **Money Clips** is geared toward consumers who want to make money, save money or invest money. **Money Clips** gives insight on everything you wanted to know about business and finance, but were afraid to ask!

- Developing your best asset: YOU
- Investing in the right mutual funds
- Low-cost loans and tax advantages
- Savings plans to build your nest egg

It's never too late (or early) to start saving and investing your money for the future. **Money Clips** is a book that will help you "get a grip" on your money matters. It provides the information that will empower you so that you can take control of your destiny. Building your net worth will give you more time for yourself or your family; it can buy better health care and child care, and it allows you the financial freedom you need to enjoy yourself—whether you're just starting out, starting over, or planning for your retirement.

Money Clips is topic-driven and organized in a way that makes it easy to understand and simple to follow. Topics include: how to assess your personal financial needs; understanding debt and other four-letter words; investing in the securities market; picking a money manager; household finances; spending, saving and retirement; and more!

Each topic is explained in detail and accompanied by cartoons, charts, graphs, quotes, examples, tips and traps, and additional resources designed to help you understand the concepts in order to make educated financial decisions.

Every section features a list of frequently asked questions and provides real-life answers, as well as "Things to Think About," a list of questions you should ask before making any financial or investment decisions.

Money Clips is the perfect tool for those who want to get—and stay—financially fit!

FINANCING THE FUTURE: AN MCI LIVING CASE STUDY

An interactive two-disk CD-ROM set featuring MCI Communications, Inc.

This program is currently being used at more than 20 business schools in 12 countries. Some of the participating colleges are University of Pennsylvania (Wharton), University of Chicago, University of Virginia (Darden), University of Michigan, Ohio State University, CUNY (City University of New York), Georgetown (Washington, D.C.), Southern Methodist University (Texas), London Business School (England), INSEAD (France), Tel Aviv University (Israel), University of Science and Technology (Hong Kong), and University of Western Ontario (Canada).

The case objective:

▲ To develop an understanding of the complex nature of corporate financial decisions.

▲ To gain insight into the highly regulated telecommunications industry.

▲ To explore financial alternatives facing a fast-growing company during volatile economic times.

Weaving a richly detailed database that covers more than 30 years of information on the economy, capital markets, the telecommunications industry, FCC regulations and worldwide events, with an in-depth analysis of telephony giant MCI, this two-disc CD-ROM set recreates the circumstances surrounding MCI as it struggled for survival a decade ago.

This interactive, multimedia case study empowers each user to participate in the analysis and decision-making process to develop alternative courses of action.

Finally, an entertaining and informational multimedia program is available for business students, professionals and executive training programs. With hours of video and audio clips and well-designed graphics and slide presentations, this program brings to life one of America's greatest stories of business achievement, in order to enhance the research and learning process. The Living Case: It's 1983 and you've been hired by MCI's management to develop the right financial strategy. The company needs one billion dollars in order to build out its fiber optic network and go head to head with the world's largest corporate monopoly—AT&T. In order to make your decision, this two-disc CD-ROM set includes 30 years of general and MCI-specific background information you need on:

▲ Society (world events and popular culture)

▲ Economy (GDP, CPI, M2)

▲ Capital Markets (equity market valuations, S&P 500 Index, securities issuance and interest rates)

▲ Telecommunications Industry

▲ FCC Regulations

▲ MCI's Financial Data and Forecasts

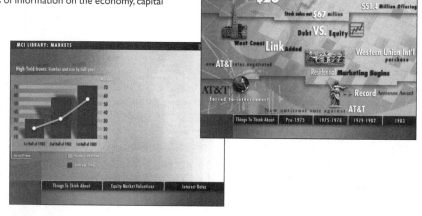

CD-ROM

These are some of the advantages of the CD-ROM learning medium:

▲ Large storage capacity of data on CD-ROMs allows depth and breadth of content coverage.

▲ Re-creation of the drama and realism of the events with the use of video, audio, photos, animation, graphs and charts, including extensive interviews with industry experts and MCI's founder and management.

▲ Easy-to-use interface captures the attention, with superior learning outcomes.

▲ Voice and pictures reinforce text as learning mechanisms.

▲ Professionals and academics can customize their learning/teaching approaches, inside and outside the classroom.

▲ A 500-word glossary provides clear definitions of the technical words used in the program.

▲ Financial information can be downloaded into Excel spreadsheets.

▲ Beta-tested at top business schools, including University of Pennsylvania (Wharton), University of Chicago, University of Virginia (Darden), Northwestern University (Kellogg) and University of Michigan. The CD-ROM program was beta-tested for more than three years and a formal marketing research firm was commissioned to perform educational testing services at UCLA with M.B.A. students.

Other Supporting Materials Available:

Case Study
A written case study to complement the CD-ROM written by one of the most distinguished professors in finance, Professor J. Fred Weston of UCLA. The written case study is intended to parallel the multimedia version and can be employed where a school's technological constraints limit their ability to use the interactive version.

Teaching Note
There is also available a teaching note that analyzes and discusses possible solutions to the questions posed in the Weston case write-up.

Glossary
Plus, a glossary containing terms used in finance and the telecommunications industry. This is a useful guide that can be employed in conjunction with either version of the case or as a free-standing resource for users of the program.

Also available is **One-on-One**, video-taped interviews with senior management of MCI.

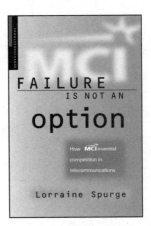

MCI: Failure Is Not an Option
How MCI Invented Competition in Telecommunications

Lorraine Spurge

ISBN: 1-888232-08-0
384 pages 6 1/4 x 9 1/4
Business/Finance
$27.95 hardcover
(In Canada: $38.95)
Rights: World

ISBN: 1-888232-41-2
$16.95 softcover
(In Canada: $23.95)

Educational and entertaining, *MCI: Failure Is Not an Option* profiles MCI's stirring history from its meager beginnings to its present success, offering an enlightening view of the financial, management, and marketing issues the company faced. Readers will experience the tension and suspense as MCI fights for survival, takes on AT&T (then the mightiest corporation in the world), shakes up federal regulatory agencies, races to raise desperately needed capital, and ultimately alters forever the American business landscape.

MCI: Failure Is Not an Option features a two-color layout sprinkled with charts, graphs, photos, and time lines illustrating the history of MCI and the telecommunications industry in general.

VIDEO/AUDIO

CASE STUDY VIDEOS AND AUDIOS

Wharton Real Estate Seminar

Running Time:
Tape 1 1 hour, 31 minutes
Tape 2 1 hour, 32 minutes
Tape 3 1 hour, 49 minutes
Includes a companion booklet

Dr. Peter Linneman, director of the Wharton Real Estate Center (ranked as the Number One real estate department according to *U.S. News & World Report*), hosts a three-part seminar with leading financial and business professionals. This seminar is an invaluable guide for any potential real estate investor or professional who wants to know how to survive—and even prosper—in the turbulent real estate market. Participants frankly share how to find and create value in the 1990s and beyond.

The three tapes encompass:
Session 1   The Future of Real Estate: Contrasting Perspectives
Session 2   Retail and Entertainment: Competing with the Home
Session 3   The Changing Demands of Real Estate Capital

The Merton Miller/Michael Milken Debate

Running Time:
1 hour, 10 minutes
Includes a companion booklet

The ultimate bus

Theory Meets Practice! Join **Nobel laureate Merton Miller** and **financier Michael Milken** for a stimulating debate about such important financial issues as:

▲ Does Capital Structure Matter?
▲ Are Markets Efficient?
▲ How Does Regulation Affect Business and Markets?

One-on-One

Michael Milken interviews the world's leading corporate entrepreneurs

MCI Communications: CEO Bert Roberts and past and present executives, William Conway, Jr., Laurence Harris and Wayne English, tell the story of how David took on Goliath in the telecommunications industry, and won!

Running Time:
1 hour
Includes a companion booklet

McCaw Cellular: Founder Craig McCaw and corporate executives Don Guthrie and John Stanton track the company's evolution from start-up to corporate giant, to its merger with AT&T.

Running Time:
1 hour
Includes a companion booklet

BUSINESS SOLUTIONS SYSTEM

Introducing the
Business Solutions System

As part of Knowledge Exchange's commitment to producing and disseminating the most useful and beneficial business knowledge available, we are proud to launch our new Business Solutions System. In order to provide a full range of products, this System will include practical, comprehensive, full-color, illustrated encyclopedias, dictionaries, and industry and trade books that cover all aspects of eight critical business disciplines.

REFERENCE ESSENTIALS SERIES

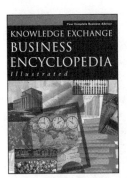

Knowledge Exchange Business Encyclopedia
Your Complete Business Advisor
Lorraine Spurge, Editor in Chief
ISBN: 1-888232-05-6
750 pages 7 1/2 x 10 1/2
Business/Reference
$45.00 hardcover
(In Canada: $54.00)
Rights: World

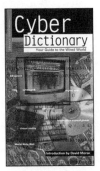

CyberDictionary
Your Guide to the Wired World
Edited and Introduced by David Morse
ISBN: 1-888232-04-8
336 pages 5 1/4 x 9 1/4
Reference/Computer
$17.95 softcover
(In Canada: $21.95)
Rights: World

Knowledge Exchange Business Encyclopedia, an exciting new resource, provides hundreds of solutions to everyday business problems. Throughout its 750 information-packed pages you'll find authoritative definitions, detailed charts and graphs, illustrations, mini case studies and much more! The result of a collaboration among some of today's finest business minds, this reference contains valuable information from academics at such prestigious institutions as Carnegie Mellon, Harvard Business School, UCLA and The Wharton School.

- An ideal reference tool to define and resolve everyday business problems, your employees—from managers to support staff—will want to keep it handy and refer to it often
- Useful as a teaching guide during formal training sessions
- Executive assistants should keep one within arm's reach to answer questions they would otherwise have to ask their bosses

CyberDictionary is the book our Net-obsessed world has been waiting for—an indispensable guide to the wide-open frontier of cyberspace.

Organized in an A-to-Z format, *CyberDictionary* is the book for the tens of millions of Net surfers and computer users. This reader-friendly, full-color reference book is packed with more than 900 easy-to-understand definitions of cyberwords, mini essays, and time-saving tips to make navigating the Net easy, exciting, and enlightening.

A Chronology of Business Poster
Limited Edition!
Available only from Knowledge Exchange, the full-color, exquisitely detailed *A Chronology of Business* poster (22"x 40") is based on the time line featured in the *Knowledge Exchange Business Encyclopedia*, which traces the history of business from its roots in 3000 B.C. to the present.
$25.00

Knowledge Exchange Business Encyclopedia Canvas Tote Bag
Vibrantly colored, heavy-duty, natural canvas, oversized tote bag (14"h x 16"w x 9"d) displaying the Knowledge Exchange logo, which is surrounded by a sampling of terms and references cited in the *Knowledge Exchange Business Encyclopedia*.
$50.00

BUSINESS SOLUTIONS SYSTEM

MANAGEMENT CONSULTANT SERIES

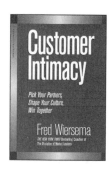

Customer Intimacy
Pick Your Partners, Shape Your Culture, Win Together

Fred Wiersema

ISBN: 1-888232-00-5
240 pages 6 1/4 x 9 1/4
Business/Marketing
$22.95 hardcover
(In Canada: $27.95)
Rights: World

ISBN: 1-888232-00-5
$14.95 softcover
(In Canada: $19.95)
Audiobook
ISBN: 1-888232-01-3
$14.00 (In Canada: $17.00)
90 minutes/Read by the author

One in three market-leading companies attains prominence today by making the most of what author Fred Wiersema calls "customer intimacy." This engaging book reveals why the most successful businesses are those that build close win-win relationships with their customers.

Richly illustrated with examples of some of the best-known and most successful customer-intimate businesses, *Customer Intimacy* is for companies wondering what to do next after having exhausted the potential of quality thinking, lean management, and business reengineering.

Fad-Free Management
The Six Principles That Drive Successful Companies and Their Leaders

Richard Hamermesh

ISBN: 1-888232-20-X
208 pages 6 1/4 x 9 1/4
Business/Management
$24.95 hardcover
(In Canada: $29.95)
Rights: World

The business place has become saturated with quick fixes that promise faster, better products and happier, more loyal employees. Unfortunately, however, these fads often waste time and energy. In this new book, Richard Hamermesh argues against this trend and stresses the necessity of getting back to basics.

Readers of *Fad-Free Management* will reap the benefits of the knowledge Hamermesh gained as a professor at the world-renowned Harvard Business School.

The Pursuit of Prime
Maximize Your Company's Success with the Adizes Program

Ichak Adizes, Ph.D.

ISBN: 1-888232-22-6
304 pages 6 1/4 x 9 1/4
Business/Management
$24.95 hardcover
(In Canada: $29.95)
Rights: World

Companies, like people, follow definite growth stages—infancy, childhood, adolescence, and *prime*. It is in this last stage of development that both humans and companies are at their best. In *The Pursuit of Prime*, Ichak Adizes, Ph.D., provides a step-by-step guide for helping businesses reach this pinnacle of corporate life.

The Pursuit of Prime provides case studies of successful companies such as Bank of America and the Body Shop and enumerates the bad habits, philosophies, and myths that prevent companies from becoming attuned to their life cycles and thereby prosperous.

The Tao of Coaching
Boost Your Effectiveness by Inspiring Those Around You

Max Landsberg

ISBN: 1-888232-34-X
200 pages 6 1/4 x 9 1/4
Business/Management
$22.95 hardcover
Rights: U.S.

Get the most out of your human capital—your employees—by transforming them into all-star managers and team players. Ideally, managers should be coaches who enhance the performance and learning abilities of others. They must provide feedback, motivation, and a master game plan.

In *The Tao of Coaching* Max Landsberg shares his belief that managers must possess a broad repertoire of management styles. The coaching skills managers can acquire from reading this book will allow them to diagnose different employee styles and use appropriate means to bring out the best in all individuals with whom they work.

BUSINESS SOLUTIONS SYSTEM

ENTREPRENEURIAL ADVISOR SERIES

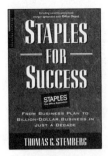

Staples for Success
From Business Plan to Billion-Dollar Business in Just a Decade

Thomas G. Stemberg

ISBN: 1-888232-24-2
192 pages   6 1/4 x 9 1/4
Business
$22.95 hardcover
(In Canada: $27.95)
Rights: World

Audiobook
ISBN: 1-888232-25-0
$12.00 (In Canada: $15.00)
60 minutes/Read by Campbell Scott

This engaging story details Staples' birth and subsequent transformation into office-superstore giant. Stemberg's hard work and commitment to excellence turned a radically simple idea into the $11 billion office-superstore industry we know today. The Staples story stands as a guide to forward thinking and successful management from genesis to innovation, to large-scale, almost limitless growth.

Staples for Success is a must-read for every entrepreneur and anyone who believes in a great idea.

The World On Time
The 11 Management Principles That Made FedEx an Overnight Sensation

James C. Wetherbe

ISBN: 1-888232-06-4
200 pages   6 1/4 x 9 1/4
Business/Management
$22.95 hardcover
(In Canada: $27.95)
Rights: World

Audiobook
ISBN: 1-888232-07-2
$12.00 (In Canada: $15.00)
90 minutes/Read by the author

The World On Time is the inspirational story of how Federal Express became a leader in the overnight-delivery industry. James C. Wetherbe, a preeminent business consultant and academic, provides a richly detailed, intimate portrait of Federal Express. Readers will learn how eleven innovative management strategies employed by Federal Express have set the standard for the way companies manage time and information, plan logistics, and serve customers.

INDUSTRY EXPERT SERIES: HEALTH CARE

Changing Health Care
Creating Tomorrow's Winning Health Enterprise Today

Ken Jennings, Ph.D., Kurt Miller, and Sharyn Materna of Andersen Consulting

ISBN: 1-888232-18-8
336 pages   6 1/4 x 9 1/4
Business/Health Care
$24.95 hardcover
(In Canada: $29.95)
Rights: World

One of the major health care tasks is to deliver more value to consumers through better and expanded products and services. *Changing Health Care* outlines the strategies that all health care organizations must adopt if they want to regain their competitive edge.

The authors propose eight winning strategies designed to keep health care providers on the cutting edge—Keep Ahead of Consumers; Keep the Promise; Cut to the Moment of Value; Mind the Cycle of Life; Capitalize on Knowledge; Ride the Technology Wave; Give Your Best, Virtualize the Rest; and Mine the Riches of Outcomes.

Prescription for the Future
How the Technology Revolution Is Changing the Pulse of Global Health Care

Gwendolyn B. Moore, David A. Rey, and John D. Rollins of Andersen Consulting

ISBN: 1-888232-10-2
200 pages   6 1/4 x 9 1/4
Business/Health Care
$24.95 hardcover
(In Canada: $29.95)
Rights: World

Audiobook
ISBN: 1-888232-11-0
$12.00 (In Canada: $15.00)
60 minutes/Read by the authors

Authored by leading experts from the world's largest consulting firm, *Prescription for the Future* profiles an industry undergoing transformation and offers insights into the challenges facing the health care industry as it employs new technologies. This book directly addresses the concerns of managers and professionals in the health care industry regarding rapidly advancing information technology, which creates both new freedoms and new problems.

Spurge Ink! and Knowledge Exchange
Price List

	BOOK	Order #	AUDIO	Order #
SPURGE INK! TITLES				
MCI: Failure Is Not an Option (hardcover)	$ 27.95	080	NA	
MCI: Failure Is Not an Option (softcover)	$ 16.95	412	NA	
Money Clips (softcover)	$ 13.95	447	NA	
REFERENCE ESSENTIALS SERIES				
Knowledge Exchange Business Encyclopedia (hardcover)	$ 45.00	056	NA	
CyberDictionary (softcover)	$ 17.95	048	NA	
MANAGEMENT CONSULTANT SERIES				
Customer Intimacy (hardcover)	$ 22.95	005	$14.95	013
Customer Intimacy (softcover)	$ 14.95	420		
Fad-Free Management (hardcover)	$ 24.95	20X	NA	
The Pursuit of Prime (hardcover)	$ 24.95	226	NA	
The Tao of Coaching (hardcover)	$ 22.95	34X	NA	
ENTREPRENEURIAL ADVISOR SERIES				
Staples for Success (hardcover)	$ 22.95	242	$14.95	250
The World On Time (hardcover)	$ 22.95	064	$14.95	072
INDUSTRY EXPERT SERIES: HEALTH CARE				
Changing Health Care (hardcover)	$ 24.95	188	NA	
Prescription for the Future (hardcover)	$ 24.95	102	$14.95	110
CD-ROM				
Financing the Future: An MCI Living Case Study				
Two-Disk Set	$ 150.00	027	NA	

	VIDEO	Order #	AUDIO	Order #
VIDEO/AUDIO TAPES				
Wharton Real Estate Seminar				
Video/audiotapes (3) and booklet set	$ 150.00	003	$60.00	176
Volumes 1, 2 and 3	$ 60.00 ea.	178, 180, 182	$25.00 ea	177, 179, 181
The Merton Miller/Michael Milken Debate				
Videotape or audiotape and booklet	$ 195.00	164	$80.00	172
One-on-One Interview Series				
Video/audiotapes (2) and booklet set	$ 150.00	171	$60.00	184
McCaw Cellular	$ 100.00	156	$40.00	168
MCI Communications	$ 100.00	170	$40.00	169

All orders are non-returnable.
Knowledge Exchange products are available wherever books are sold.
Spurge Ink! is the exclusive representative of Knowledge Exchange books, audiobooks and multimedia products.
Bulk Orders: Call for purchase order information and corporate price breaks.

For more information contact:
Spurge Ink!
16350 Ventura Boulevard, Suite 362
Encino, California 91436
Tel: 800.854.6239 or 818.705.3740
Fax: 818.708.8764
Visit us on the Internet @ www.spurgeink.com

Spurge Ink!
Making Business Accessible